25 Home Automation Projects for the Evil Genius

Evil Genius Series

Bionics for the Evil Genius: 25 Build-it-Yourself Projects

Electronic Circuits for the Evil Genius: 57 Lessons with Projects

Electronic Gadgets for the Evil Genius: 28 Build-it-Yourself Projects

Electronic Games for the Evil Genius

Electronic Sensors for the Evil Genius: 54 Electrifying Projects

50 Awesome Auto Projects for the Evil Genius

50 Model Rocket Projects for the Evil Genius

Mechatronics for the Evil Genius: 25 Build-it-Yourself Projects

MORE Electronic Gadgets for the Evil Genius: 40 NEW Build-it-Yourself Projects

101 Spy Gadgets for the Evil Genius

123 PIC® Microcontroller Experiments for the Evil Genius

123 Robotics Experiments for the Evil Genius

PC Mods for the Evil Genius

Solar Energy Projects for the Evil Genius

25 Home Automation Projects for the Evil Genius

25 Home Automation Projects for the Evil Genius

JERRI L. LEDFORD

New York Chicago San Francisco Lisbon
London Madrid Mexico City Milan New Delhi
San Juan Seoul Singapore Sydney Toronto

The McGraw-Hill Companies

1 2 3 4 5 6 7 8 9 0 QPD/QPD 0 1 2 1 0 9 8 7 6

ISBN-13: 978-0-07-147757-4
ISBN-10: 0-07-147757-8

The sponsoring editor for this book was Judy Bass and the production supervisor was Pamela A. Pelton. It was set in Times Ten by Keyword Group Ltd. The art director for the cover was Anthony Landi.

Printed and bound by Quebecor/Dubuque.

This book is printed on acid-free paper.

About the Author

A freelance writer with more than 10 years' experience, Jerri L. Ledford has contributed more than 700 articles to publications such as *Intelligent Enterprise, Network World,* and *Information Security.* In addition to covering all aspects of business technology, she writes and teaches consumer technology courses at both Hewlett-Packard and CNET. A resident of Magee, Mississippi, Ms. Ledford has also written several consumer books focusing on a variety of technology issues.

Contents

25 Home Automation Projects for the Evil Genius

Introduction to Home Automation

If you're at all interested in home automation, you've probably heard about Bill Gates' house. The Microsoft software mogul lives in a fully-automated home that's more than 6,000 square feet. It's large. And it has amenities like heated floors, digital, virtual art screens that change the pictures on display, and of course, personal climate control zones, even for guests.

The Gates' home is a picture of modern automation. The definition for home automation is the use of technology to make life more comfortable. The concept is to use technology – wiring, computer modules, and software programs – to control your lights, appliances, environment, audio, video, grounds maintenance, and dozens of other features. And of course, all of this takes place without too much thought or input on your part.

Gates' home is monitored and managed from the Redmond Campus of Microsoft. But you don't have to have an entire software company to control your home. Nor do you have to have the billions of dollars that Gates invested to automate your home. All that's required is this book, a few supplies, some tools, and the ability to follow directions.

So what's home automation all about? The computers and wiring that make it possible. Many people assume that you have to be a master software developer or a licensed electrician to do home automation on your own, but that's not true. Home automation projects can take anywhere from 30 seconds to several months to complete. Most of what's covered here is the shorter variety.

These projects also don't require a deep knowledge of architecture or wiring. But don't let that mislead you. There are some home automation projects that require a professional. Those projects are more about home integration, or integrating the various home automation projects that you'll learn about here (and more) into a single system.

Wiring

Wiring for home automation can take several forms. For example, you can add automation to your home during the new construction phase of home building. This is by far the easiest way to add home automation to your living space because contractors can handle all of the planning and wiring, and even the configuration of your home automation applications and technologies.

Not everyone is building a new home, however. For people who have an existing home, then you can add home automation to your current wiring scheme, by connecting the home automation devices and technologies to your existing wiring. This isn't as difficult as it sounds. As you'll learn a little later in this chapter, basic wiring requires just a few tools and is fairly easy to do.

The decision to add home automation to a new-construction home or to an existing home is an easy one – either you're building a new home, or you're not. But there's another decision that's a little more difficult to make. It's the decision to make your automation projects wired or wireless.

Not all home automation projects can be wireless. However, anything that's related to your audio and video components probably can be. Additionally, there are some computer-based home automation projects that can also be made wireless. There are advantages and disadvantages to both configurations. If you choose to wire all of your home automation projects into your existing wiring, you could find that the wiring is too old and will need to be replaced. This is especially true if your home is more than about 20 years old.

Going wireless eases the frustrations of having to run new wiring and it makes it faster to integrate home automation into your house, but it isn't all roses. When you choose to go wireless with your home automation projects, you run the risk of having a wireless router or access point fail. When that happens, you have no control over your automation controls.

Computers

Going wireless means having a computer. In fact, there are many home automation projects that you can control with computers. Bill Gates has enough computer-driven automation capabilities that all of his systems are monitored on the Redmond Campus of Microsoft. This ensures there is someone watching over his systems all the time.

You can't be Gates, but you can have computer-controlled automation. Just don't try it with an older-model computer. Instead, invest in a new model computer that's operating with Microsoft Windows XP or Windows Vista when it is released.

The updated operating system will ensure that your computer can control the most automation projects. It also ensures that you have the most automation options possible available to you.

There's more about home automation networking later in this chapter, so keep reading.

Remote Access

The final consideration as you add home automation is how you plan to access your home automation once it's installed. Again, you have several options:

- **Internet-Based Access**: This type of remote access is enabled using the Internet. It means that your home automation projects must be connected to a router or other device that gives you access to the controls over the internet. This is great for some home automation features, but not practical for others.
- **Remote Control Access**: Remote access via remote controls is probably the type of access with which you're most familiar. You can choose from infrared remotes, much like the one you use with your television, or radio frequency which is available over longer distances. Radio frequency remotes are most often associated with garage door openers.
- **Automatic Controls**: Of course, you can set some of your home automation features to take place automatically. For example, you can set your lighting to increase or decrease at certain times of day. You can do the same thing with your thermostat. These aren't true forms of home automation in the sense of integration, but they're still nice to have.

Networking

Networking is where you get deep into the theory of home automation. Depending on who you talk to, true home automation involves networking all of your automated functions into a single network. However, unless you truly are Bill Gates, this may be more involved than you want to get with your home automation projects.

Most users want to network together the audio and video – entertainment – portions of their home automation but networking other functions isn't as important. The good news is, networking some functions isn't as difficult as you might believe. The bad news is that there are functions you may want networked that will require a professional.

You'll find networking specifics in the appropriate chapters. For now, brush up on your networking knowledge. You'll probably need it later on.

Equipment Needed

I'm sure you're getting anxious to get started, but there's one last thing to cover before we get into the home automation projects – equipment. Each of these automation projects that are included in this book will have very specific requirements, and each will differ depending on the project. But there are a few tools that you should plan to have around for nearly everything that you do. Those tools include:

- Screw Drivers (both Phillips and flat heads)
- Wire Strippers/Wire Cutters

- Pliers (both needle nose pliers and crimper-stripper pliers)
- Circuit Tester

These are just your basic tools, so you'll also find a complete list of tools and equipment needed at the beginning of each project.

Let's Get Started

Home automation includes a lot of theory and various skills but if you don't get too hung up in the minutia of concepts you can complete a home automation project in a short amount of time. In fact, automating some functions will take you less than half an hour to do.

I know you're anxious to get started, so let's not waste any more of your time. Let's jump right into the first of your projects.

Project 1

Lights Please: Indoor Lighting

Equipment Needed

- Circuit Tester
- Flathead Screwdriver
- Phillip's Head Screwdriver
- Pliers
- Wire Cutter/Wire Stripper
- Wire Nuts
- A ZWave Lamp Module
- A ZWave Dimmer Switch
- A ZWave Remote Control

Understanding Lighting Control

In home automation projects, lighting is one of the simplest, and easiest to install projects. It's also one of the most convenient home automation projects that you can do. In older homes, especially, the light switches and controls are often located in inconvenient places, so switching lights on and off means more effort from you. If you can control those lights with a remote, you never have to leave the easy-chair to turn off the overhead that's blinding you.

With some lighting systems, automating the control of your lights is as simple as plugging a module into the wall, connecting a lamp to it, and configuring a remote control. Of course, lighting automation can be much more complicated than that, as well, but the more complicated jobs are generally saved for new construction homes because they require special in-wall wiring that's difficult to implement (but not impossible) in existing homes.

When it comes to automating your interior lighting, you have several different options that run the gamut from inexpensive and specific controls to comprehensive – and expensive – controls. In general, there are three types of lighting control systems:

- **Architectural Systems**: Architectural systems are lighting control systems that operate lights in your entire house. These types of systems require a network of low-voltage cabling that runs through the entire house, and therefore are usually only used in new construction houses. Architectural control systems can be used in existing houses, however, the house will have to be rewired to support this type of system. Either way, architectural controls systems are the most expensive of the three. This type of system can also control draperies, audio and video components, and other home automation features, but they usually require professional installation.
- **Power-Line Controlled Systems**: These types of control systems use the existing house wiring and can be as simple as dimmer switches or as complicated as networked lighting systems that are controlled by a keypad in each room. This is the type of lighting control that we'll use in this automation project.
- **Radio Frequency Controlled Systems**: The difference between power-line controlled systems and radio frequency controlled systems is the way in which they are controlled. Radio frequency controlled systems use radio waves to control lighting.

One of the fastest (and easiest) ways to automate your lighting is using a system designed specifically for this purpose. Project 3 is all about the X-10 system, but before we get there, let's start with something a little easier to install. There are several systems that fit the bill, including the Insteon SwitchLink system and Sylvania's ZWave system.

We'll focus on the ZWave system, though all lighting automation systems work in about the same way. Each system has both wall switches and plug-in modules that allow you to control not only your lights, but other appliances as well. In addition to these, some systems have keypads and all of them have remote controls that give you additional options for controlling lights or appliances once you have them connected.

The ZWave system is a radio-frequency controlled system, so there is no need to install additional routing equipment or to rewire your house. However, a USB-enabled ZWave controller is available should you want to control your lights using your computer or home network.

> **Note**
>
> The ZWave system is not compatible with the X-10 system, so if you plan to install an X-10 system in your house, you may want to explore the lighting options in that system before you begin installation of the ZWave system.

Terminology

Before we get into the details of installing components of the ZWave system, let's look at some of the terminology you might want to know.

Grouping: A grouping is a set of lights that you want to control all at one time. For example, if you want to set your lights so that you can turn all of them on in one room with the push of a button, then these specific lights or fixtures are grouped together as a single unit.

Scene: A scene is a special lighting scenario that you can create with the push of a button. For example, maybe when you come home in the evening you want the table lamps in your living room all on full brighteness. But then later, when you sit down to watch television, you want the lamps to dim a little, and you want the accent lighting in the room to come on. These are two separate scenes that you can set up for the lighting in any given room.

Dimming: Reducing the brightness of one single or group of lights.

Figure 1-1 *The ZWave dimmer switch.*

Timers: Timers can be set to turn lights on or off at any given time. These controls are especially helpful if you plan to travel or be away from home later than usual. You can set a timer to automatically turn on certain lights throughout your house and then to turn them back off. This gives the appearance of occupancy, even when no one is home.

The Equipment

There's one more thing to look at before we get to the installation instructions – the actual ZWave modules and controllers. We won't cover the keypad in this book, but what we will look at are the switches, plug-in modules, and remote controls, shown in Figures 1-1, 1-2, and 1-3.

The three components shown are sold individually. The cost for these items runs from about $25 for each piece and up. There are also several starter kits, which include the remote and one or two modules. These can be purchased for around $75–$125. In most cases, however, the dimmer switches will have to be purchased separately.

Figure 1-2 *The ZWave lamp module.*

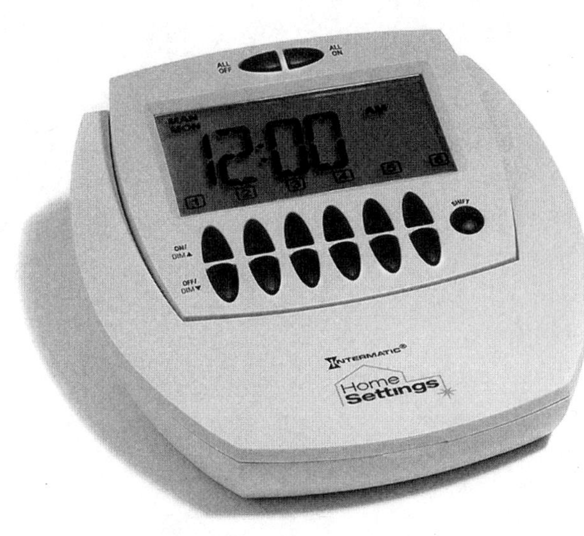

Figure 1-3 *The ZWave remote control.*

Installation Instructions

Warning

Always turn the power off at the circuit breaker before you begin working with electricity. Once the power is off, test your outlet or receptacle with a circuit tester before beginning to work with the wiring. Also be sure to wear the proper safety equipment during any home automation project. Basic safety gear includes goggles and breathing protection.

Installing Lamp Modules

Installing the lamp modules for the ZWave system is really as simple as plugging the module into the wall socket and then plugging the lamp into the module. That's all there is to it. Once the lamp is plugged in, there is a button on the module which you can use to turn the lamp on and off. You can also dim the light by holding the button in for a few seconds.

This button is also used to add a specific lamp to a group or scene setting (which you'll learn more about below in the Programming Your Remote section).

Installing the Dimmer Switch

Installing the dimmer switch is a little more involved than installing the lamp modules, but in all, it's still not a difficult process. For this installation, you'll need to have your circuit tester and both flathead and Phillip's head screwdrivers available. You may also want to have a pencil and paper or masking tape and a pen available to create a diagram of your wiring or to label the wires so you'll know how to connect the dimmer switch.

1. Begin by shutting off the power to the switch at the breaker box. If all of your circuits are properly labeled, this should be easy. Just locate the correct circuit breaker and push it into the off position. Some electricians, though, aren't altogether organized, so it's possible that the circuit breakers in your breaker box are either incorrectly labeled, or worse, not labeled at all. If this is the case, you'll need to either use a circuit finder, like the one shown in Figure 1-4 to locate the correct circuit, or you'll need to turn each circuit off in turn until you locate the one you're looking for.

There are pros and cons to both approaches. The circuit breaker finder will cost you about $25–$50 but it saves on the guess work. If it's

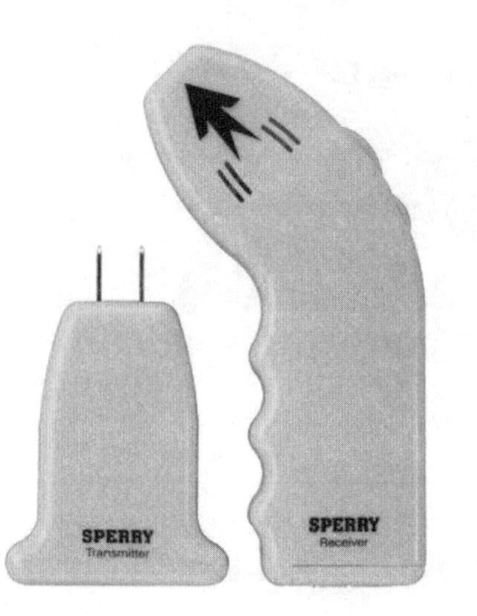

Figure 1-4 *Use the Sperry Circuit Breaker Finder to locate unlabeled circuits.*

money you prefer to save, the only real downfall with turning each circuit breaker off until you find the right one is that you'll probably have to reset every digital clock in the house when you're finished.

2. Remove the two screws from the switch plate, and then carefully pull the switch plate away from the wall. As Figure 1-5 illustrates, the light switch, which is recessed into an electrical box in the wall, will then be exposed.

   Before you go any further, be sure to test the wires attached to the switch to ensure that there is no electrical current flowing through them. Even if you think you've broken the current to the switch, it's still possible that the switch is still electrified or that there's incorrect wiring that allows current to continue to flow to the switch once the circuit is thrown. **Always test wires for current before you begin working with them**.

3. Again, being very careful, disconnect the switch from the mounting box to which it is attached by removing any visible screws. You may have to gently pull the switch out of the electrical box to disconnect the wires. Make note of which wire is attached to what connection. In general, black wires are positive and carry the current to the switch from the power source. White wires are usually neutral, and they complete the electrical circuit. Green wires or wires that have no insulation covering them are usually ground wires. These wires help reduce the risk of electrical shock.

Figure 1-5 *Remove the switch plate to expose the switch and connections.*

   In addition to these wires, you may also find a red wire attached to the switch, especially if the switch is one in a series or if there are multiple switches wired into the same receptacle. The red wire is also positive, and it's sometimes called a traveler wire, because current travels through it to the next switch or outlet in a series.

4. Now, connect the new ZWave dimmer switch by connecting the wires to the new dimmer switch according to the manufacturer's directions. Figure 1-6 shows how it might look. The switch will either have screws under which the wires should be secured, or it will have wires that must be connected to the existing wiring. If this is the case, hold the matching wires (black with black, white with white, and green with green, etc.) with the exposed ends parallel. Then slip both ends into the opening in a wire nut and twist the wire nut until the wires are securely bound together.

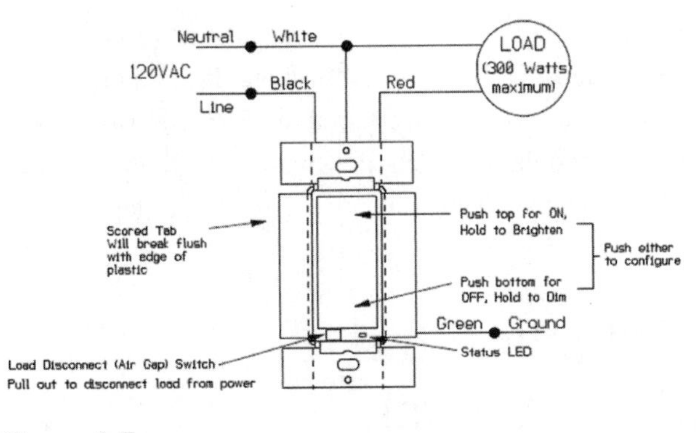

Figure 1-6

If you have additional wiring that's not being used, be sure to cap it off with a wire nut. **Never leave exposed wires uncapped**. Uncapped wires can cause dangerous conditions that lead to fires or electrocution.

If necessary, bend the tabs on each side of the switch with pliers until they break loose so that it fits properly into the switch box. Set the switch into the box and screw into place.

Screw the new faceplate into place over the switch. Your new switch should come with a new faceplate, but if it does not, you can use the old faceplate as long as it fits. Due to the difference in designs, however, you may find that the old faceplate doesn't fit. If one hasn't been included with the new switch you can probably find one that does fit at your local hardware store.

Once you have the switch installed, restore the power to the switch, and test it. Pressing the top of the switch should turn the light on, the bottom should turn it off, and if you want to brighten or dim the lights, press and hold the top or bottom of the switch until light reaches the desired level.

Configuring the Remote

When you get all of the modules and switches installed, then it's time to program your remote control. The remote may have time and date settings on it. If this is the case, use the manufacturer's instructions for creating those setting.

Creating a Network

It doesn't matter if you have a single lamp module or several modules and a switch, you need to create a network through which you can control your lights. Basically, a network lets you assign modules to the controller that you plan to use. All modules must be assigned to the master controller before you can program or control them using the remote.

1. On the master remote, press and release the **INCLUDE** button. The **INCL** light should begin to flash, indicating that the controller is ready to sense and include other modules or switches you want to control.
2. While the **INCL** light on the remote is flashing, press and release the **PROGRAM** button on the module or switch that you want to control. When the connection is made between the module or switch and the remote, the **INCL** light will stop flashing and the world **SUCCESSFUL** will show on the control screen for about 2 seconds.
3. If the phrase **NOT SUCCESSFUL** appears on the screen, try linking the controller with the module or switch again. Once successful, move on to the next module or switch that you want to use your remote to control. You can have more than 50 devices or switches on one remote (the exact number is determined by the system that you choose) and the pairing process must be completed for each device or switch you want to include.

Create a Lighting Scene

Now that you have created a network of the lights that you want to control, you can set each light to the specific scene, or lighting level, that you prefer. This requires that you first associate each switch or module with a channel and then set the lighting level that suites you.

1. On the remote controller, simultaneously press and hold the channel **On** and **Off** buttons until the world **LEARN** is displayed. Once **Learn** appears on the screen, release the buttons.
2. Now press and hold the **INCLUDE** button on the remote controller. When **INCL** appears on the screen, press the **PROGRAM** button on the *lighting module or switch that you wish to program*, and then adjust the light to the desired level. Once the light level is set, release the **INCLUDE** button. Now your lighting scene is set.

There's much more to the ZWave system, but this is a good start. And since the system is easy to use, you'll be controlling not only your lights, but also some appliances, in no time at all. Just follow the manufacturer's instructions that are included with the remotes, modules, and switches.

Troubleshooting

If only every project could go smoothly without any problems cropping up. Some do, but others just don't, no matter how careful you are. Here are a few of the most common problems that you're likely to encounter during and after the installation of these lighting control systems.

My light worked before I put the new switch in, but it doesn't now. What's wrong.

It might seem a little simplistic, but did you reconnect the circuit at the breaker box? Sometimes in our hurry to check out our new handiwork, we forget the smallest of details. No current passing through the circuit is sure to prevent the new switch from working.

If you have reconnected the circuit breaker, then it's possible that you mis-wired the switch. Throw the circuit breaker again, remove the faceplate from the switch and double check your wiring against the manufacturer's instructions.

I plugged in the module, but now my lamp won't work.

Did you plug the lamp into the module? The lamp must be connected to the module to receive current to operate it. If you did plug the lamp into the module, and it's still not working, check the circuit breaker. It's possible that you threw the circuit while connecting the module and the lamp.

If the circuit is still active, then it's possible there is either a switch controlling the outlet where the module and lamp are plugged in or that you have a bad controller module. Check the switch first, and then if necessary, exchange the module for a new one.

I pressed the **INCLUDE** button on my remote, but I can't get the remote to recognize the module or switch.

Was the **INCL** light still flashing when you pressed the **PROGRAM** button on the module or switch? If not, start over, making sure to press the program button while the **INCL** light is still flashing. If the light was still flashing when you pressed the **PROGRAM** button on the module or switch, then walk through the process again, just to be sure you completed all the steps.

Selecting the Right Switch

A switch is a switch, right? Well, not exactly. There are several things to consider when picking out the right switch for your home automation project. For example, do you need a low-voltage switch or a dry-contact switch? And which is best suited to the project, a toggle switch, a dimmer switch, or momentary switch?

All of these elements are determined by the project that you're working on, but there are a few general guidelines that will help you pick the right switch. Before we get that far, though, let's discuss what each type of switch is, exactly.

- **Low Voltage Switches**: Low voltage switches basically work by changing the amount of power that flows through the wiring connected to the circuit. The change in power is then read by a switch, or piece of equipment, and specific actions take place in response to the voltage changes. You may have heard these switches referred to as digital sensors.
- **Dry-Contact Switches**: If you have any regular light switches in your home, then you've seen dry-contact switches before. These switches operate by opening or closing a connection to a circuit which effectively allows or disallows the electricity to travel through the connected wires. When the switch is turned on, electricity travels through the wires to the fixture or device being controlled. When the switch is turned off, no current can reach the fixture or device.

Low voltage and dry-contact switches are usually designed by a manufacturer. For example, a dimmer switch is a low-voltage switch. The amount of current that flows through the dimmer switch is determined by the voltage allowed at the switch. Normal toggle light switches, like the ones commonly found in most homes are dry-contact switches.

Other switch considerations:

- **Toggle Switch**: A toggle switch is either off or on, there are no settings in-between. Toggle switches are useful if you know that you'll either need the power to a specific fixture or device turned on or off at any given time. Most of the existing switches in your home are probably toggle switches unless your home has been built in the last decade. In which case, you may have some dimmer switches installed.
- **Dimmer Switch**: Dimmer switches are low voltage switches that allow you to adjust the fixture or appliance that is controlled by the switch. Most often these switches are used for lighting and for fans, but you may find them in other uses, too. Dimmer switches can also be used with remote controls to control the level of the lighting (or other fixture in a room) via an infrared signal.
- **Momentary Switch**: A momentary switch is one that causes an action and then automatically returns to the state it was in before the switch was activated. For example, a doorbell is a momentary switch. When you press the doorbell button, the switch is activated, but when you release the button it is deactivated. These switches are good for doorbells (of course) as well as garage doors.

So, how do you know what's the right switch for the automation project that you're doing? In most cases, there will be some literature for the project that you're working on that will tell you which type of switch that you'll need. However, if there is nothing that tells you, you can usually determine the type of switch that's needed by the project that you're doing. For example, if you're installing a light that you want to be able to adjust to suit your lighting tastes, a dimmer switch is your best option. Or if you're installing a garage door, then you might consider the momentary switch.

Standard light switches, called toggle switches, can be used for any fixture that you install that needs only to have power on or power off.

One more thing that you might want to take into consideration when you're installing a new switch is the switch plate. There are two things to consider when replacing your switch plate. The first is the size. When you pull the old switch off the wall, you may find that there is a gaping hole behind it that the switch plate is hiding.

If this is the case, you'll want to be sure that the switch plate you choose is the right size to cover the hole. And even when holes in the wall aren't the problem, you may find a paint ring around the old switch plate or you may be adding some additional switches to the same plate, so you'll need a larger plate. Of course, there's also a space factor, and in some cases there just isn't room for larger switch plates.

Be sure to measure carefully **before** you purchase the switch plate to save yourself a trip back to the hardware store if the switch plate that you've selected doesn't fit.

The other consideration with your switch plate is the color. Two standard colors for switch plates are white and a tan-ish or yellow-ish off-white. These colors should be easy to find, but you can even find a variety of patterns or even blank switch plates that you can paint or cover on your own so they'll match your décor.

A good rule of thumb when choosing a switch plate that does fit with your décor is to choose one that's either the color of the trim in the room or one that matches (or contrasts) the color of the paint or wallpaper on the walls. Just don't choose something that contrasts too starkly or you'll find that the switch plate becomes the focus of the room instead of blending nicely into your décor.

Who would have thought that choosing the right light switch could be such a chore? It's not really. It's just a few simple decisions that you need to keep in mind so you don't find yourself with a switch that's not appropriate for the project you have in mind or a switch that looks completely out of place in the room you're project is designed to enhance.

Project 2

Intruder Alert: Outdoor Lighting

Equipment Needed

- Motion Sensor Light Fixture
- Screwdriver
- Wire Nuts
- Masking Tape for Labeling Wirings
- Circuit Tester

Installing Motion Sensor Lighting Outdoors

Motion sensor lighting is one of those useful home automation projects that not only makes life a little simpler, but they also help to cut down on the cost of electricity. The way a motion sensor light works is that it casts a set of infrared beams in a specific pattern around it. When someone enters the pattern, the beams are broken so the lights come on. If there's no activity to break the beams for a specified amount of time – usually 1 to 5 minutes – the lights gradually dim down to nothing.

This is particularly handy if you come home after dark and have forgotten to turn your outdoor lights on. The motion sensor lights will turn on when you enter the range defined as you set it up, and then they'll turn off when there's no longer activity in the sensor zone.

The security aspect of these lights is also convenient. No need to worry if someone is in or around your house. These lights are activated by motion, so anyone walking into the sensor zone is greeted by the light coming on. Intruders prefer the cover of night, so they're not likely to stick around once the lights come on.

Motion sensor lights come in a variety of types from X-10 enabled to integrate with your X-10 system (if you have one installed) to fixture replacements, like the one shown in Figure 2-1. You can even get motion sensors that simply screw into an existing receptacle and solar-powered motion sensor light.

Figure 2-1 *A fixture replacement motion sensor light.*

To install a motion sensor light outdoors, you first need to find the right fixture. These modules can be found at Radio Shack, Home Depot, Lowes, or any other home improvement store. You can also find them online by searching for the name of the switch that you want to install. Froogle.com is a good way to quickly locate the least expensive products.

1. Turn off the power to the fixture at the circuit breaker.
2. Remove the old fixture carefully. Unscrew the base and gently pull the fixture away from the wall.
3. Use the circuit tester to check that there is no electrical current passing through the wires that should now be exposed.
4. Carefully disconnect the wiring to the old fixture by unscrewing them from the connections or by removing the wire nuts holding the fixture wires to the house wiring, as shown in Figure 2-2. As you're disconnecting the old fixture, be sure to mark the wiring according to function. You can do this by either drawing a diagram of the fixture wiring and label it or by marking each wire with labeled masking tape.
5. Discard old fixture.

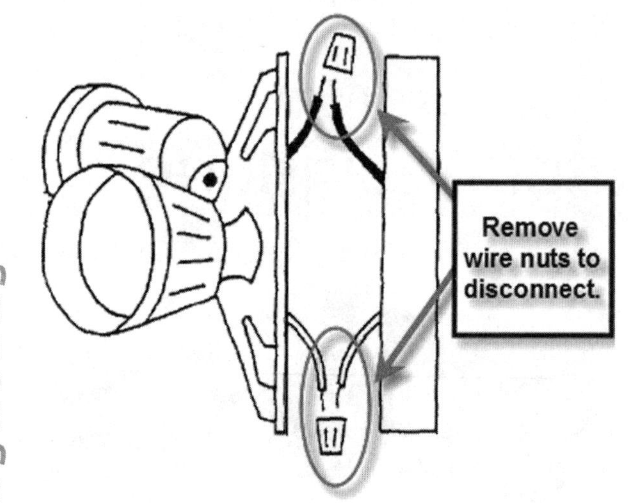

Figure 2-2 *Remove the wire nuts to disconnect the existing fixture.*

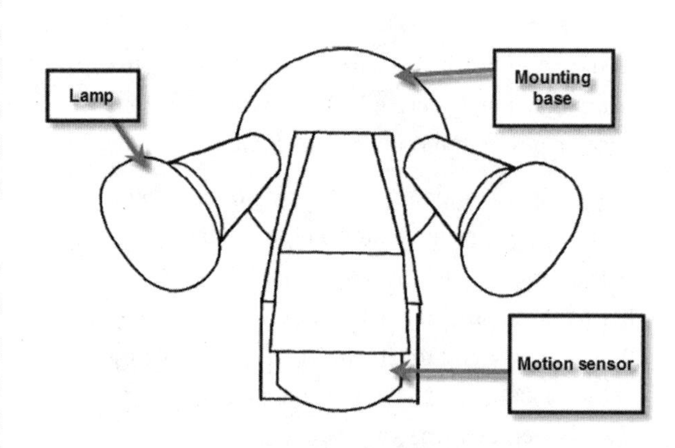

Figure 2-3 *Sample motion sensor light fixture design.*

6. Install new fixture by connecting wiring according to package directions. A simple diagram of a motion sensor light is shown in Figure 2-3.
7. When connecting the new fixture be sure that wires are matched according to function: hot to hot, ground to ground, and line to line.
8. Attach new fixture to the old fixture plate by aligning screw holes and securing in place with new screws.

Once your fixture is in place, then you need to test and adjust the sensitivity of the motion sensor.

1. On the bottom of the fixture there should be a button labeled "Test." Slide this button into the test position. Then set the sensitivity level to "medium."
2. Reconnect the circuit by switching the circuit breaker to the on position.
3. Turn on the main power to the light fixture. The light will come on. If there's no movement in the detection zone, the light will remain on for about 30 seconds before going out. Any movement in the detection zone will cause the light to remain on.
4. If you have an adjustable detection zone, readjust the sensor of the monitor to the area in which you want the light activated.
5. Be sure the lights in the fixture don't interfere with the sensors. Try to position so that the lights and/or bulbs are at least ½ inch away from the sensor.
6. Adjust the sensitivity so that lights come on when you want them to, and test this setting by walking through the sensor zone at the longest point from the sensor. A sample sensor zone is shown in Figure 2-4. The zone will vary some according to the design of the motion sensor light that you've chosen to install. However, this diagram should help you to visualize the sensor zone as you're setting the sensitivity.
7. When sensor zone and sensitivity are set, slide the switch from the 'Test' position to an active position. The light should now come on when it senses motion and go off after motion has stopped for several seconds.

At some point, you may decide that you want to change the sensitivity or sensor zone for the light. This is an easy to accomplish task.

1. Carefully move sensors to cover the desired zone. If no one is in the zone, the setting should be correct when the light does not come on.

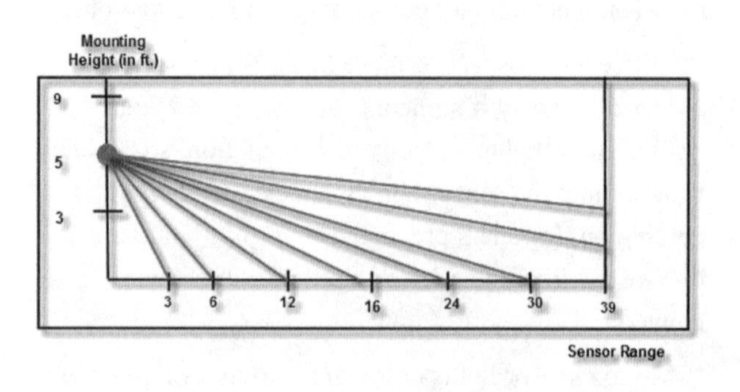

Figure 2-4 *A sample motion sensor zone.*

2. If you can't keep the sensor from picking up unwanted motion, you can use black electrical tape to cover portions of or spots of the lens to prevent it from picking up motion.

Troubleshooting

I've installed the light, but it never goes off. What can I do?

If you've installed the light, but it never seems to go off, check the sensor zone or sensitivity level to which the light is set. If necessary, you can block portions of the sensor with black electrical tape to prevent it from picking up unwanted motion.

I've installed the light, but it never comes on. What can I do?

If you've installed the light, but it isn't activated by motion, it's possible that you've installed the light improperly. Try disconnecting and reconnecting the fixture using the manufacturer's directions. If the light still doesn't seem to be working, it's possible that it's defective. You may want to exchange it for a new fixture.

The light seems to flicker on and off all the time. How do I stop it?

The problem might be that there's not enough motion or not enough regular motion to keep the light on. You may want to consider resetting the sensitivity or duration of the light to improve the problem. If this still doesn't work, then you should consider replacing the motion sensor light with a timed light to ensure that the bulbs stay lit for the duration desired.

Additional Security Monitoring Sensors

Automated outdoor lighting is a blessing. When you forget to turn the porch light on before you leave (and then return after dark) you don't have to worry that you won't be able to see to get the key in the lock. Automated outdoor lighting also increases the security of your home.

You can set your outdoor lighting to automatically come on when something or someone crosses the sensor path. Then, if an intruder tries to gain entry into your house after dark, the lights will automatically come on, which is usually enough to scare the intruder away. Thieves prefer to work under the cover of darkness. They're not interested in getting into your home if they know the whole neighborhood can see them doing it.

In addition to outdoor lighting, there are other types of security sensors that you can wire into your home automation system to help protect you and your family, too. For example, did you know that you can set up motion sensors that don't trigger lights, but will trigger an event such as the barking dog alarm that we'll discuss later in the book?

You can also set up sensors that will enable an event if there is glass breaking in your house, or if there is smoke or carbon monoxide in the air.

All of these sensors are available from two places. You can either hire a security company to come wire your house for security, or you can install and program the sensors yourself.

If you choose to have a security company come take care of creating your security system, you'll also have the option of having that company monitoring your alarms. A monitored alarm system can reduce the amount of time that you spend worrying about someone breaking into your home or that a fire will destroy your home. As soon as the alarm monitoring company receives an alert that something is wrong, they'll contact the proper authorities and dispatch police or firemen if necessary.

The downside is that you will likely end up paying a monthly fee for the privilege of having someone else monitor that alarm system. That fee will run you anywhere from $50 to $250 depending on the service you contract for and the area in which you live.

If you choose to install the program sensors yourself and monitor them with your computer then you have many options for security sensors. Here's a quick list of the sensors that are available to you:

- **Magnetic Sensors**: These sensors are most commonly used for doors and windows and operate on magnetic contacts. As long as the contacts are connected, the alarm is set. When the contacts are broken, your alarm will sound. In most cases, these are hard-wired security sensors, but some wireless sensors can also be found. These units connect back to the central alarm monitoring device.

- **Floor Sensors**: Floor sensors can either emit infrared beams or be weight controlled. The infrared sensors sit at floor level and when the beam is interrupted the alarm sounds. Weight sensors are usually installed under floor coverings and are activated when the weight load on the sensor changes (for example, when someone steps on or near them). Bill Gates' house is equipped with floor sensors that monitor where guests are. That information is then used to track guests and to ensure they are as comfortable as possible during the time they are in the house.
- **Glass Sensors**: These sensors are used to detect breaking glass. If someone is trying to break into you home by breaking the glass to get to the lock, the sensor is tripped and the alarm is sounded. Glass breaking sensors can be used on windows, doors, and other glass openings that need to be monitored.
- **Motion Sensors (also called Motion Detectors)**: Just as the name implies, these sensors detect motion in a given area. And motion sensors can be programmed to detect small sized movements (for animals) or larger movements (for people). The signal that is sent when a motion sensor is activated can then be used to trigger a variety of actions from alarms going off, to lights being turned on, or sounds being played.
- **Vehicle Sensors**: Vehicle sensors monitor when vehicles arrive in the area that the sensor is monitoring. You might use a vehicle sensor to control a gate to your home, or even an alarm that lets you know when a vehicle has pulled up in front of your house. You probably won't want to use a vehicle sensor in the average neighborhood unless your house is situated far enough from the road that it won't detect the normal flow of traffic.
- **Environmental Sensors**: In some areas of the country, flooding and high winds or high temperatures are a concern. Water, wind, and temperature sensors – also called environmental sensors – can be used to monitor the environmental conditions and programmed to alert you if an abnormal condition should arise. For example, if you live in an area that's prone to flooding, then you can place a water sensor in a specific area and if that area begins to flood you'll be alerted of the rising waters.
- **Electrical Sensors**: Electrical sensors monitor the flow of electricity into your home. In most cases, these sensors are only used by business (such as data centers and hospitals) where electricity is a concern. However, when installed in your home and connected to a monitoring program that sends alerts to you via telephone or e-mail, electrical sensors can let you know if the power to your home is interrupted for any reason.
- **Smoke and Carbon Monoxide Sensors**: These environmental sensors monitor the quality of the air in your home and alert you when air quality is affected by smoke or carbon monoxide. When the air quality drops below a certain point, an audible alarm is sounded to let you know there's a problem. If you're monitoring your home via the Internet with management software, then you will be alerted via e-mail or telephone.

Installing your own alarm/home monitoring system can be a grand undertaking. If this is an option that you would like to consider, take the time to learn how these sensors and the wiring and management systems that go along with them work before installing the system. You can find more information about them by searching the Web or by purchasing books from Amazon.com or your favorite bookseller.

Moving On

Motion sensor lighting is typically used outdoors, but it can be used indoors. It's especially useful in rooms that aren't used for long periods of time or aren't used frequently. For example, a motion sensor light works well in closets, in the laundry room, or in stairwells where you may have your hands full when you're moving through those areas. They can be used in other areas, as well. It all depends on how you use your home.

No matter how you use the motion sensor lighting, however, it's not hard to install them. In fact, depending on the type of motion sensor lighting that you use, it might be much simpler than what's covered here. It could also be far more complicated, should you decide to use a whole house lighting control system.

This information is designed to get you started. From here, you can imagine the different possibilities for lighting controls inside your home, so it's time to move on to lighting controls outside the home. The next project can walk you through that.

Project 3

Whole House Automation: Installing an X-10 System

Equipment Needed

- Home Control Assistant 6.1 Plus
- X-10 Lamp Module

Anytime you look at home automation equipment, you'll probably also see the term X-10. X-10 is one of the most popular ways to automate a home because – aside from the fact that X-10 is more prevalent than mosquitoes in August – the equipment is inexpensive, and very easy to install. What's more, those components are often portable, meaning that you can automate your home, even if you live in an apartment or other rental property. When you move, simply uninstall the components and take them with you.

So, what exactly is an X-10 system? Well, X10 is technically not a system so much as it's a communications language (also called a protocol) that allows you to control lights, appliances, and other equipment in your house via the existing electrical wiring. Costly rewiring of your home is unnecessary – a major benefit that saves you both time and money. The protocol is the basis of design for components that can "speak" to each other (i.e. a remote can communicate with a light switch or an appliance using the X-10 protocol).

To control lights and equipment, an X-10 system uses commands that are sent via the controller, whether it's a remote or an in-wall control module. Each switch or receiver is designed to receive these commands.

There are several components in an X-10 system, though in most cases, you can have a single piece, such as a single lamp or light switch, or a whole houseful of receivers that are controlled by a single controller. And that single controller can be a remote control, the telephone, or even your PC.

Here are the major components of a whole house X-10 system:

- **Wireless Controllers**: Controllers are the equipment that sends out a command to control the receiver modules. There are a variety of controllers available from universal transmitters (such as remotes) and timers to computer interfaces and telephone responders.
- **Transceivers**: X-10 systems communicate via radio frequency (RF) signals. A transceiver converts RF signals to X-10 signals so that modules and wireless controllers can communicate.
- **X-10 Receiver Modules**: These are`your control modules for lights and equipment throughout your house. The most frequently used receiver modules within the X-10 system are plug-in modules, switches, and micro modules.
  The modules receive the commands from the controllers via RF technology.

Installing an X-10 Computer Interface

Most home automation projects start as a disparate collection of X-10 modules scattered throughout your house. And that's one of the great benefits of the X-10 system. You can start small and build up over time. But it's also possible to install a computer program that makes managing multiple modules much simpler. The program provides a single control center from which you can manage everything from the address (or location) of the module to any time-controlled commands. And when you combine the computer interface with a remote networking application, it's feasible that you could control your home from anywhere in the world.

We're not going to go quite that far, but in the pages that follow, you will learn how to install and activate a Home Control Assistant (HCA) 6.1 Plus designed by Advanced Quonset Technology, Inc. The HCA is a piece of software that you can purchase at most home improvement stores and online from web sites such as SmartHome.com.

Installing the Software

The Home Control Assistant is prepackaged software. Once you've purchased the software, the first thing you need to do is install it on your computer.

> **Note**
>
> This text assumes that you are familiar with computers and have a basic understanding of how to use them, as well as how to install software. If you do not know these things, you should take a beginning computer course before attempting this project.

Once you've installed the software it's time to set up a new home design. A home design is like a plan for how you'll automate your home. The program should have installed an icon on your task bar during setup. Click that icon to access the program.

If you don't see an icon for the program in your task bar, you can also access it by going to **Start > Programs** and looking for the Home Control Assistant program name.

> **Note**
>
> In order for the HCA software to operate any X-10 enabled devices, it must be running on your computer all the time. This also means that your computer needs to be turned on all the time. Should the program become disabled, or if your computer is shut down, the software will not operate.

When you open the software for the first time, you'll be taken to an overview screen. The first thing you should do is begin a new home design by clicking **File > New.** You'll be prompted to select either **Quick Start Wizard** or **New Home Wizard.**

- **Quick Start Wizard**: This wizard begins with a tutorial and then guides you through each room of your home making suggestions where automation might be used. It also creates a complete home design including existing (or new) devices, programs for device operation, groups of devices that you want to work together to create a scene, schedules for specific tasks, and control displays to help you navigate controls when accessing the program from your computer.
- **New Home Wizard**: This wizard is much simpler than the Quick Start Wizard. It creates an empty design to which you can add devices, schedules, programs, and groups using additional HCA wizards. This is the option we'll use for this exercise.

First, you must supply a name for your design. This name appears in the HCA title bar when this design is loaded. You can create several designs, each with a unique name, and in different files. This is an especially useful feature if you choose to experiment with more than one configuration for your programs and schedules.

Next, you'll be asked to choose your location to determine sunrise and sunset times for your schedules. You'll be prompted to enter the location of your home by latitude and longitude to achieve the most accurate location. However, if you don't know your latitude and longitude, you can also select from a list of states and major cities. You really don't need to be too exact, as the time difference between the sunrise and sunset from your longitude and latitude and the nearest city should only be a few minutes.

The New Home Wizard

The New Home Wizard helps you set up the new file for your home. The program creates an empty design to which you can add your devices, programs, schedules, and groups.

After you create the name and time zone for the program, you'll be taken to a screen that looks like the

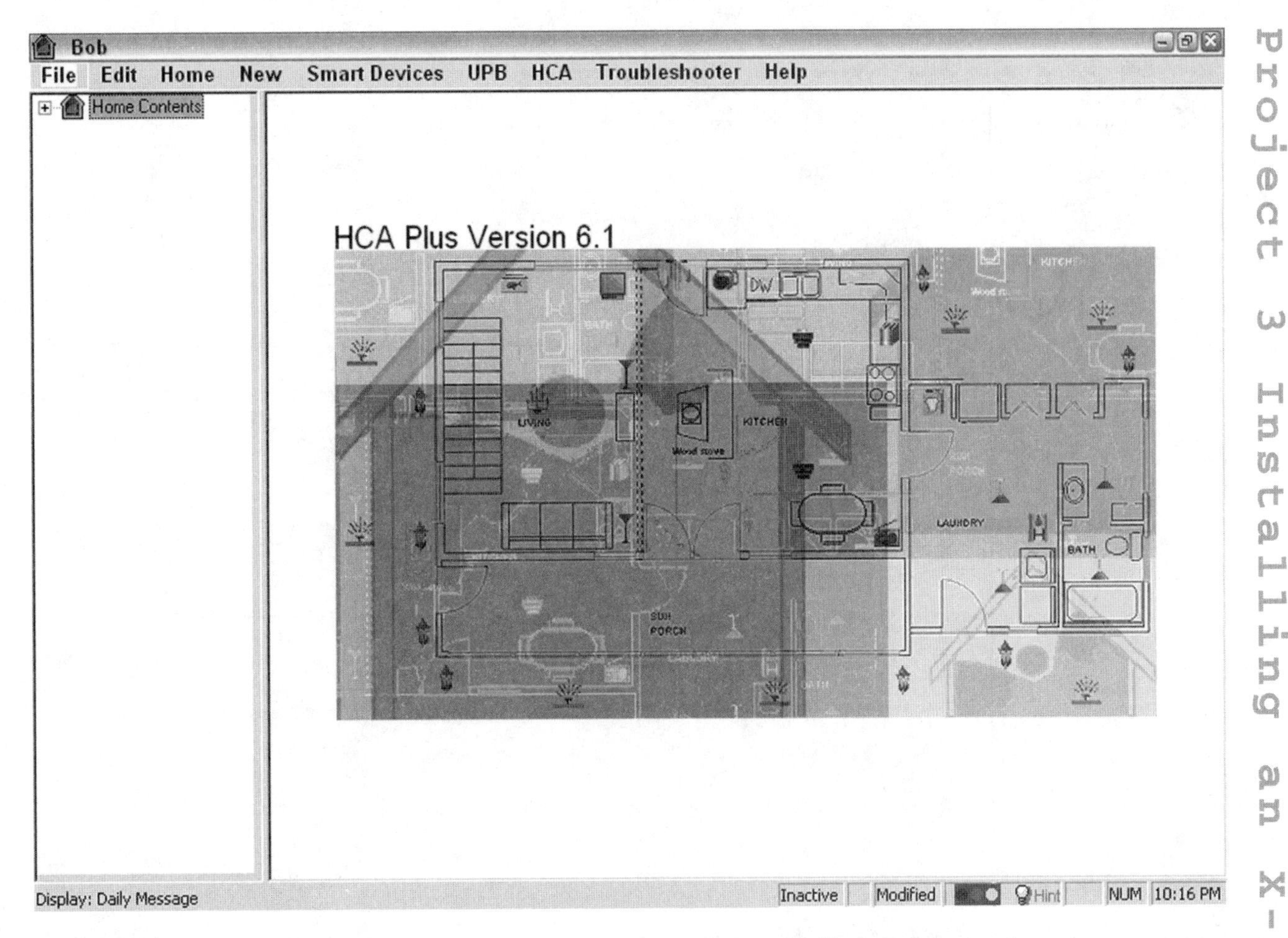

Figure 3-1 *When you start a new home, the program creates and empty profile for you to populate.*

one shown in Figure 3-1. This is just a blank slate to which you can add all of your modules and programs.

Because the New Home Wizard creates an empty design, the next step is to add a display with a floor plan background, device, schedule, and schedule entry to create a sample home.

1. Go to **New** and click **Display**. The New Display Wizard appears, as shown in Figure 3-2.
2. Enter a name for your display and select the design folder in which you want it stored and then click **Next**.
3. Select what will appear in your display and then under Background select **DXF file**.
4. Browse to the location on your hard drive where the HCA files are stored and **myhome.dxf**. Be sure Broderbund 3D Home Architect 2.X is selected in the **What program produced this file** text box and then click **Next**.

Figure 3-2 *Use the New Display Wizard to set the properties of your display.*

5. On the next screen, select **Copy the DXF image from its file to the HCA file** and click **Next**.
6. Select the desired margin size on the next screen and click **Next**.

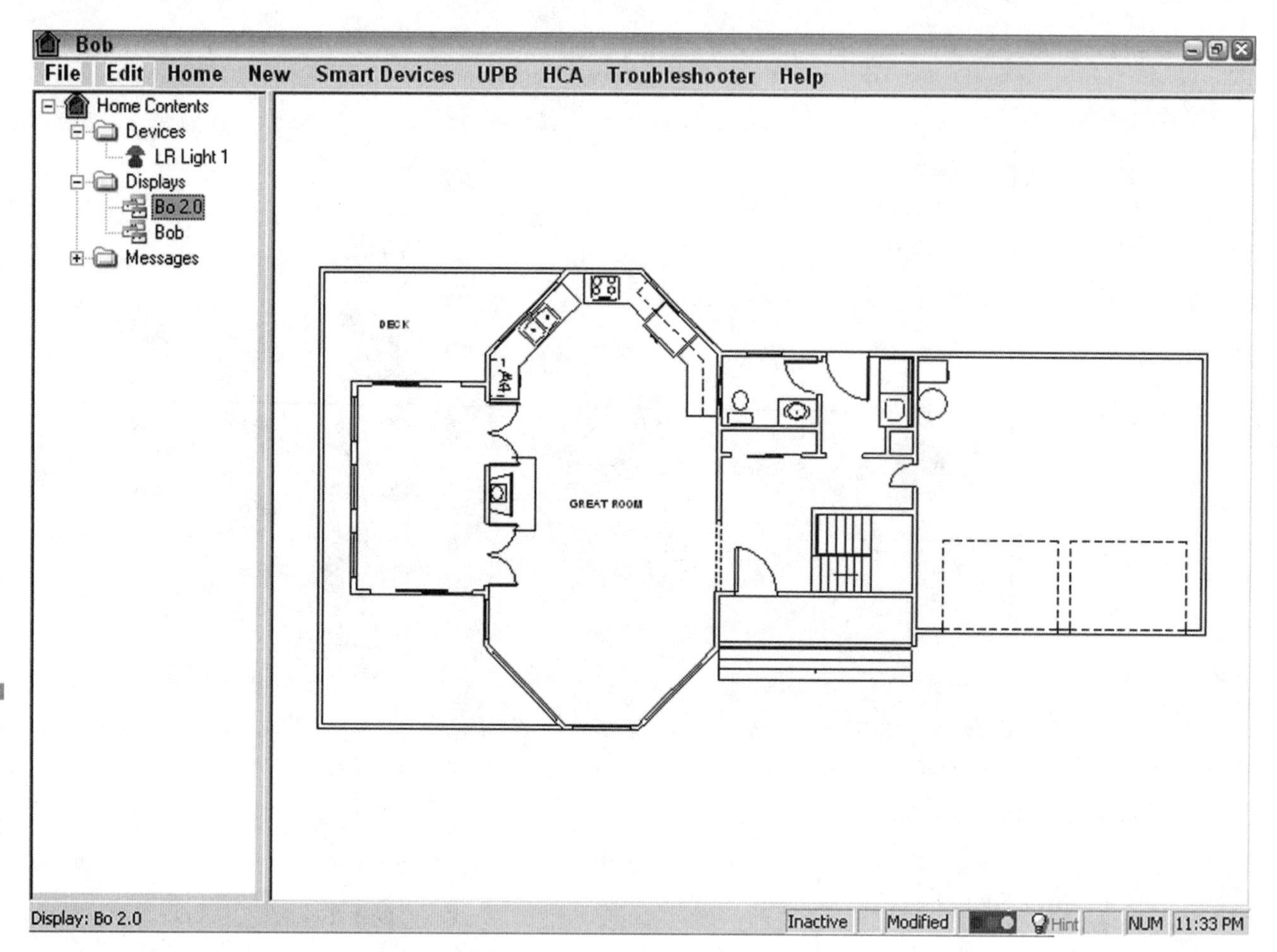

Figure 3-3 *Use the sample floor plan to add devices, groups, programs, and schedules.*

7. Now you need to choose your drawing layers. Select each item that starts with **FL2**, and move that item to the list on the right by clicking the right arrow button on the wizard.
8. Click **Finish**.

As shown in Figure 3-3, you can now see a sample floor plan to which you can add devices, groups, programs, and schedules.

Create a Device

With your floor plan in place, you can now add devices to your display.

1. From the main menu, click **New** and select **Device**. The **New Device Wizard** begins.
2. In step one, type a name for the device that you're adding. Then select **Displays** for the design folder and select the device type that you want to add. Don't worry if your device isn't listed. Just use the closest match. It won't interfere with the program at all.
3. In Step 2, enter any notes that you might find useful at a later time. It's not essential that you enter anything on this screen, so if you don't have any notes, just click **Next** to move to the next step.
4. In Step 3, select your light module and then click **Next**. Again, it's not essential that your parts match exactly, just select the closest part listed.
5. In Steps 4, 5, and 6, leave all of the settings as they are and click **Next** to bypass each screen.
6. In Step 6 of the wizard, use the right arrow button to move the select display from the **Displays** list to the **Put icon on these displays** list.
7. Click **Finish**. Your icon will probably be located in some weird place on your floor plan. Click and drag the icon to place it where you want it.

Now it's time to connect the lamp. First, you have to connect a lamp to a lamp module (plug the module into the wall and plug the lamp into the module). Assign the setting of the module to A1.

Back at your computer, right-click on the icon you just placed on your floor plan. Use the menu that appears to turn the lamp on and off. If the lamp can be controlled by the computer, the next step is to set a schedule for turning the lamp on and off automatically.

Create a Schedule for the Device

With the lamp you've just added to your HCA network working properly, it's time to create an automatic schedule for it. This allows the HCA system to control the light without any input from you.

1. In HCA click **New** and then select **Schedule**.
2. In the Schedule Wizard, type a name for the schedule you're creating. For example, you might choose to label your schedule vacation (for those times when you're out of town). Once you've named your schedule, click **Finish**.

Set a Schedule

Once you've created a schedule, then you need to add entries to the schedule for controlling your devices.

From the main toolbar, click **New** and then select **Schedule Entry**. The **Schedule Entry Wizard** launches.

1. In the first step of the wizard, select the device and schedule that you want to work with from the dropdown menus. If desired, you can add a name for the schedule entry (such as TV Time). Then click **Next**.
2. In Step 2, select **Schedules both the time the device goes on and when it goes off?** Then click **Next**.
3. On the next screen, Step 3, select the times when your device should go on automatically. Then click **Next**.
4. Next schedule the times when you would like the device to go on. Click **Next**.
5. In Step 5, select the illumination level you want the lamp to come on at and click **Next**.
6. Now set the time you want your device to go off. Then click **Next**.
7. The final step is to review the schedule entry you've just created. When you're finished reviewing it, click **Finish**.

That's really all there is to it. You now have an X-10 system control that will control all of the modules that you have throughout your house, right from the comfort of your computer.

Of course, there are other options in the application, too. You can add additional devices and set up groups, too. They're all controlled by wizards, so they're easy tasks to accomplish and change.

Just make sure you monitor the application for a couple of days after setting it up, however, to be sure that everything is operating the way that you expect.

Note

You must activate HCA before it does anything. When HCA is active the current schedule is watched to see when things should turn on or off. To activate the program, select **Home > Set Current Schedule** from the main toolbar. Select the schedule you want to be the current schedule and click **OK**. The status bar displays the schedule that HCA is now watching. Once you've selected the current schedule, go to the toolbar, select HCA, and then choose **Activate HCA**. That's it your schedule should be set and your modules will now operate automatically according to that schedule.

Understanding the X-10 System

When you first encounter an X-10 system, or even just the term X-10, it can be a little confusing. What exactly is X-10, and how does it work? I can hear your frustration. Let me see if I can help a little.

For starters, let's define X-10. X-10 is a protocol that standardizes power line communication among the devices that are used for home automation. Confused yet? Basically, X-10 is the standard that defines how devices, such as appliances and lighting, communicate through your house wiring.

Here's a good way to think of it. If you're at all familiar with the Internet, then you know that it's a network of interconnected computers that cover the globe. Now, think on a smaller scale, and apply that image to the electrical wiring in your house. The wires that make it possible for you to turn on a light switch and have light or to operate appliances and to plug devices into wall sockets to power them, are all interconnected or wired to the circuit box that controls the electricity for your home.

Now, that circuit box is like a hub for the internet. Signals come into that box and signals move out of that box. What comes in is the electrical current from outside your house and signals from the switches and outlets around your home. What goes out of the box is electrical current, in response to the signals that are received at the box.

So, what you have is a mini-network throughout your house that connects all of the switches and circuits together to create a cohesive whole.

"Yeah, I get all that. But what's X-10?"

Okay, so you have this network of electrical wiring, current and signals running through your home. The signals that travel from the light switch when you turn it on to the circuit box are standardized signals that all operate on the same frequency. X-10 is a protocol – or standardized "language" – that allows data to travel through the same wiring that electrical current travels through.

When you dig deep into the protocol, X-10 is a very complex mixture of signaling, timings, and encoding that enables specific commands to be transferred from one place to another in a very efficient manner. For example, say you have an X-10-enabled lamp module plugged into the wall. It takes more than just an electrical current to control that lamp with a remote control. There's some data involved, too.

X-10 makes it possible for you to send a command with the remote to the circuit box which in turn controls the amount of current that is used to control the lamp. The lamp module (which is X-10-enabled) converts the infrared control to a frequency that travels along the wire and is interpreted at the switch box.

Here's where it gets a little dicey. The lamp module (which is controlling your lamp) is assigned an address. And the circuit box executes specific commands, based on the address of that lamp module. Meanwhile, the lamp module is also 'listening' to the commands from the circuit box to know what actions to take with the lamp.

So, if you're using an X-10 remote to control your lamp, pressing a button on the remote would send a specific command to the X-10 module, which is then relayed to the circuit box, which answers with a specific action. The X-10 module is listening for this action and when it receives the command, it executes it.

Sounds like the long way around the block, doesn't it? In a way, it is. And you might see the results of that journey in the responsiveness of the devices that you're controlling with an X-10 module. It can sometimes take a couple of seconds for that command to travel the whole route, which means that your lamp won't respond to the button pressed on the remote control as quickly as your television or stereo might.

It's a short lag time, but to someone that's paying attention, it's noticeable. So, don't be surprised if your X-10 devices aren't instantly responsive. Have patience, and everything will work as it should.

There is one more thing that you should know about the term X-10. It is the name of the protocol used to allow devices to communicate between each other via the power lines in your house, but it's also the name of a company that manufactures home automation equipment. So, you may find that devices are either X-10 devices or they are X-10 enabled devices. In most cases, these devices will all work together, because they're all designed to work with the X-10 protocol.

This hasn't always been the case, but as home automation gains traction in the marketplace, fewer companies are creating X-10 modules and devices that are proprietary. Consumers like you demand that all of their devices and modules work together.

As you can see, X-10 can be a bit of a confusing term. In this book the term X-10 is used in a very generic format. Devices are X-10 enabled, modules and switches are X-10 enabled, and all of them should work together in harmony. If only the whole world could get along so harmoniously.

Moving On

Ultimately, HCA is an application that helps you gain control of your X-10 modules. There's much more you can do with X-10, but there are also plenty of books written on the topic if you're interested in learning more.

For now, you can add X-10 modules and control them from your computer. And in the next chapter you'll learn about automating your thermostat controls. Really, home automation is just this easy.

Project 4

It's Getting Hot in Here: Control Your Climate

Equipment Needed

- Screwdriver
- Wire Cutters (with Stripper)
- Circuit Tester
- PSP 711RF Digital Programmable Thermostat System by LuxPro

One area where many people prefer to have some type of automation is in their environmental controls, more specifically, in environmental heating and cooling. No one likes to be either too hot or too cold. And wouldn't it be really nice if you could just adjust the temperature of your house from the easy chair?

You can. All you have to do is install a remote-controlled thermostat. But before we get to the installation, there are a few things about thermostats that you should know.

- **Consider your scheduling needs**: Are you away from home often? Or are you home more often than you're away? If you're home often, you don't need a system that's controllable from anywhere as much as you need more control while you're home. You scheduling needs should dictate the type of climate control system you decide on.
- **Consider the location of the thermostat**: As the saying goes, "location, location, location." The location of your thermostat can make a huge differece in the comfort of your home. For example, if your thermostat is located above a light switch (and many in older homes are), it could create a difference in the thermostat reading and the actual temperature of your home. The light switch gives off heat that can skew temperature readings. If this is the case, you may want to consider a thermostat that you can move from room to room to ensure that you're comfortable in any room in your home.
- **Consider how you will want to control the thermostat most often**: This goes back to how often you'll be home. If you're always gone, you may want to access your thermostat from the computer. But if you're home about the same amount of time that you're away, a programmable thermostat could be a better choice. You just program it once and then forget about it.
- **Consider your room temperature sensors**: Room temperature sensors go hand in hand with the location of the thermostat. A moveable room temperature sensor is handy, but maybe what you really want is sensor zones – sensors in multiple rooms. These give you the comfort you desire without the need to carry a room sensor everywhere.

Depending on your heating and cooling needs (and your personal desires), you have several options for installing automated climate controls.

- **X-10 Compatible Thermostats**: Thermostats that integrate with your X-10 system for computer control.
- **Non-X-10 Compatible Thermostats**: Thermostats that are automated and can be controlled by remote or telephone, but are not X-10 integrated.
- **Automated (but not remote controlled) Thermostats**: These thermostats may or may not integrate with your X-10 system, but are not usually compatible with remote controls. These are the "program-and-forget" thermostats that do everything for you.

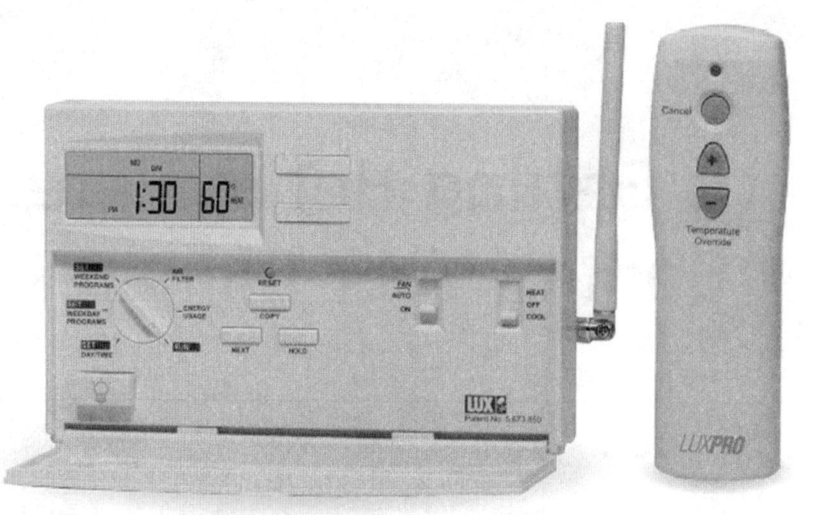

Figure 4-1 *The PSP 711RF is both programmable and remote controlled.*

In this project, we'll use the PSP 711RF Digital Programmable Thermostat System by LuxPro, shown in Figure 4-1.

Installing the Thermostat

Installation of the PSP 711RF thermostat is really very simple. It requires nothing more than replacing your existing thermostat with the PSP 711RF and programming your remote. It's a project that takes only a few minutes to do but makes it infintely easier for you to control your thermostat, without having to get up from your chair. And, because it's programmable, you have the option to set it and forget it.

Removing the Old Thermostat

The first step is to remove your old thermostat from the wall. Before you begin, remember to switch electricity to the HVAC unit off at the breaker. If you want, tape a note to the breaker box to warn others not to turn the breaker back on. This is one way that you can be assured that no one will accidently reinstate the power while you're working with the wiring.

1. Remove the cover from the old thermostat. Most thermostats have snap-on covers, like the one shown in Figure 4-2.
2. When you remove the cover, you'll see several wires and the temperature control, as shown in Figure 4-3. Before you begin, test the wires with the circuit tester to ensure there is no current running through them.
3. There should also be two or three screws holding the thermostat in place. Disconnect the wires and remove those screws and gently pull the thermostat away from the wall.

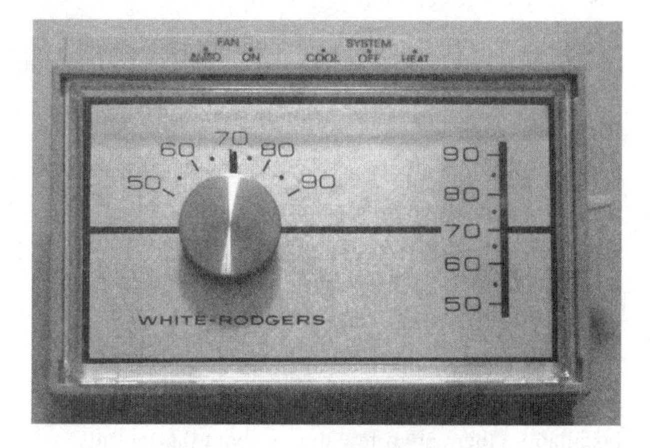

Figure 4-2 *Most existing thermostats have a removable face plate that snaps off the unit with a gentle tug.*

NOTE

In some cases, there's another plate beneath this one which must also be removed before the thermostat can be removed completely. Remove the screws for this case, and again, gently pull the plate away from the wall.

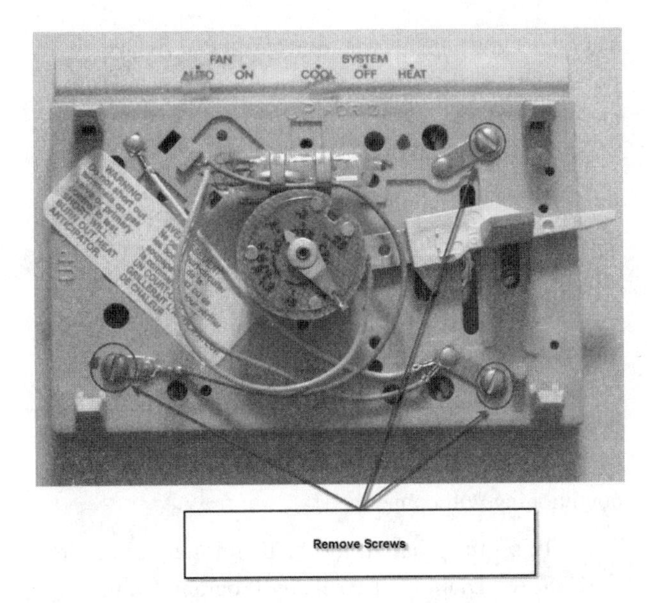

Figure 4-3 *Remove the screws holding the thermostat to the wall and disconnect wiring.*

4. As you're disconnecting wires from the existing thermostat, note the letters printed near the terminals. Label each wire for identification using tape, or draw a diagram as you disconnect the wiring. Remember to remove and label wires one at a time.

> **Caution**
>
> As you're disconnecting wires, make sure they are not so tight that they fall back into the wall if you release them. If the wire does seem snug, pull gently. If no slack is available, the wire will not move. In that case, wrap the wire around a wooden pencil to ensure it doesn't fall back into the wall.

Mounting the Wall Unit

Once the old thermostat has been removed from the wall, installing the new one is a matter of reconnecting wires and screwing the thermostat into place.

1. Before you begin rewiring the new thermostat, strip about 3/8 inch of the insulation from the wiring if needed. Also take a few minutes to clean any corrosion that might have collected on the exposed wiring. This helps ensure that your wires are connecting properly.

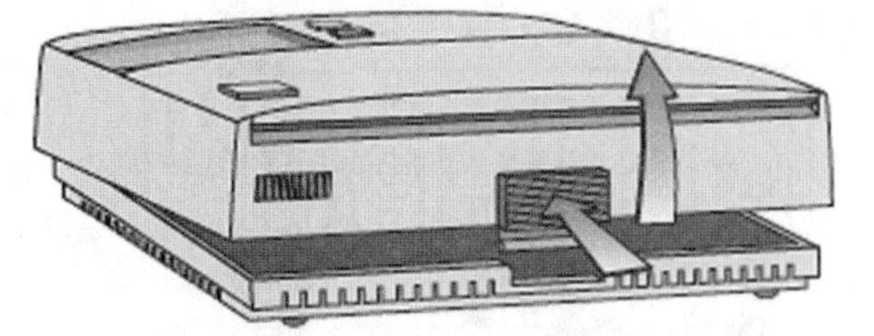

Figure 4-4 *Remove the faceplate from the thermostat before mounting.*

2. Drafts can affect the temperature reading of your thermostat, so before you place the new thermostat on the wall, take the time to fill the wall opening with a non-combustible insulation to prevent drafts.
3. Now, unpackage and prepare the new thermostat for mounting by removing the front faceplate. To do this, press up on the button on the bottom of the thermostat and swing the body away from the base and up. Gently snap the faceplate off the unit, being careful not to break the small, plastic notches that hold it in place. Figure 4-4 illustrates.
4. Thread the wiring through the holes provided in the mounting plate. Then hold the base against the wall and screw into place. If possible, use the mounting base to cover any marks from the old thermostat unit.
5. Connect the wiring to the new thermostat according to the labels that you placed on the wires during the disconnection step above. The diagram below (Figure 4-5) illustrates how your wires should connect to the new thermostat.
6. When you finish wiring the thermostat, snap the cover back on to the unit. You're now ready to install batteries in the unit and then you can reconnect your HVAC breaker and test the unit to ensure the thermostat is working.

Installing Batteries

The thermostat requires batteries to operate your HVAC unit and retain memory programming. When the batteries need to be replaced, the REPLACE indicator appears in the display. Even if the REPLACE indicator doesn't appear, you should replace the batteries in the

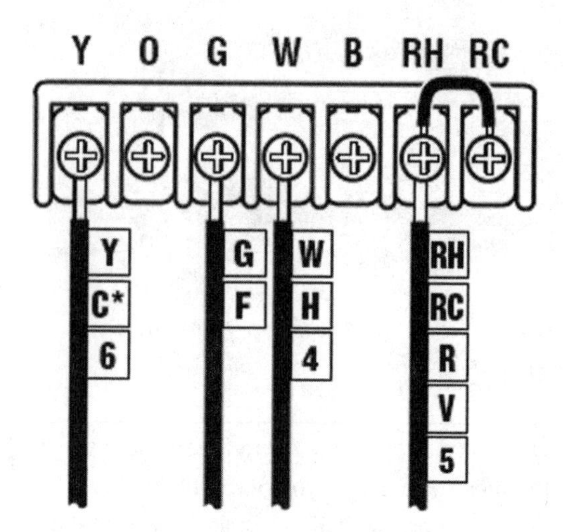

Figure 4-5 *Connect the wires to the new thermostat according to this diagram.*

unit at least once each year to ensure it continues to operate properly.

1. Remove body of thermostat as described above.
2. Remove the battery clip and batteries.
3. Install two new double A (AA) alkaline batteries in the battery compartment, making sure the batteries are installed according to the polarity marking shown in the clip.
4. Replace the thermostat body.

Now you can turn the power back on to your HVAC unit. Once it is on, the thermostat will display "SUN 12:00 AM". It will take about 90 seconds for the thermostat to begin to display the room temperature alternately with the time.

> **NOTE**
>
> When you're replacing the batteries in your thermostat, you have approximately 1 minute to complete the battery replacement without losing your existing programming. If it takes longer than 1 minute, you'll need to reprogram your thermostat to ensure it's working as desired.

Programming the Thermostat

Programming your new thermostat to automatically adjust the temperature during given times of the day is as easy as pushing a few buttons. You can change any preset times or temperatures to suit your schedules for any day of the week.

The thermostat has four periods duing any given day: morning, day, evening, and night. Each of these periods is programmed with a temperature for HEAT and a temperature for COOL. Each of the temperatures is the designated point at which the heating or cooling unit switches on. For example, if you have the morning heat point set at 68 degress and the cool point set at 78 degrees. Then when the temperature reaches 68, the heat will come on but when it reaches 78 the air conditioning will come on.

1. To set the program for the thermostat, rotate the Speed Dial to Set Weekday Program. Or, to program the weekend rotate the dial to Set Saturday or Set Weekend Program. Because most people are home more frequently during the weekend, you have the ability to set a different program for the weekend than what's in place for weekly hours, when most people are working.
2. Use the Up and Down buttons to change the start time for the morning time period.
3. When you've set the desired start time, press the Next button to set the desired temperature.
4. When you've finished setting the temperature for the morning, press the Next button again to be taken to the next time period.
5. Continue setting the time periods using the same steps you did for the morning time period and temperature settings. When you've set the program for each time of day, you can use the Copy command to copy the same settings for the remainder of the week.
6. When you're ready to set the weekend time frames, turn the Speed Dial to Set Weekend Program.
7. Once you've programmed all of the days, then turn the Speed Dial back to the Run position and your new program will be activated.

You can, at any time, change the programming of your thermostat. If you find that it's too warm or too cool, simply follow the above steps again to define a new set or programming parameters. Additionally, you have the option to temporarily change the temperature

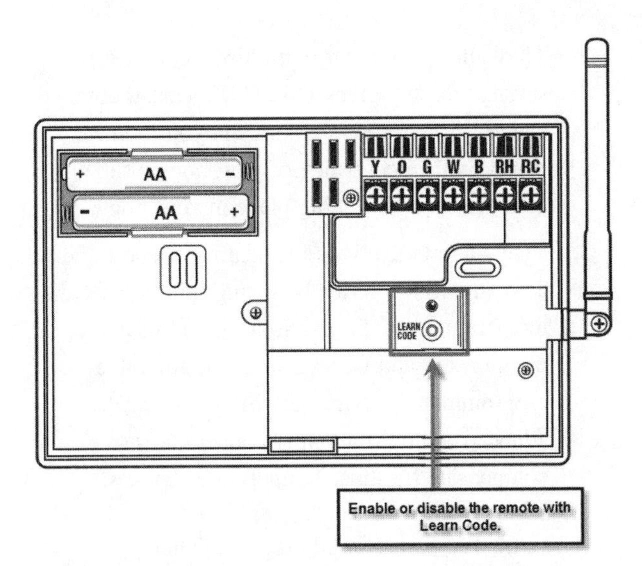

Figure 4-6 *Press and hold the learn code button until the LED light comes on to cause it to recognize the remote control.*

in your home without reprogramming the thermostat. This is accomplished by pressing the Up or Down buttons until the desired temperature is reached. The override settings will be returned to normal programming at the start of the next time period.

Finally, you can use the Hold button to change and hold the temperature without affecting your programming settings. Press the hold button and adjust the temperature. That temperature will remain dominant until you release the hold by pressing the Hold button again. This is a useful feature if you plan to be away from home for a few days and you just want to maintain a constant minimum or maximum temperature.

Programming the Remote

Now that you've got your thermostat under control, it's time to set it up to recognize your remote control.

1. Begin by removing the thermostat cover from the wall plate.
2. Press and hold the recessed Learn Code button, shown in Figure 4-6, with a small screwdriver until the LED light above the button illuminates.
3. Next, press and release any button on the remote control.
4. Wait until the LED is extinguished.
5. Replace the thermostat cover.
6. Your thermostat should now respond to commands made with the remote control. There may be a slight pause (of a few seconds) between the time you press a button on the remote and the time the thermostat responds. This is normal and should be no cause for alarm.
7. To disable the remote control, follow the instructions above, but skip Step 3. This prevents the thermostat from recognizing commands from the remote control.

Once your remote is programmed, then you should be able to use it from most places within your home to adjust the temperature, turn the fan on and off, and turn the HVAC unit on or off completely.

Changing Transmitter Security Code

If you are using more than one remote control unit in your home, the two units could possibly interfere with one another. To prevent this, make sure that each unit is using a different security code.

1. Slide the back cover off the remote control and remove the batteries from the unit.
2. In the top of the battery compartment is a three-position slide switch. This is the ID Selector. The switch has three positions. Slide it right, left, or to the center to change the ID selection. Make sure the switch is in a different location for each remote.
3. When you've changed the ID selection, replace the batteries and the battery compartment cover.
4. Once the remote is reassembled, you'll need to enable your remote again, using the Learn Mode function covered above.

After you've completed these steps, your remotes should work without interfering with each other. So, if you want to keep a remote in the bedroom and a remote in the living room or kitchen, you can use them without any difficulties (and without the need to carry a remote with you everywhere you go in the house).

Choosing the Right Thermostat

Thermostats are pretty unexciting little boxes that take up a small amount of space on the wall somewhere in your house. At best, you may only notice your thermostat when the temperature in your house isn't comfortable. At worst, it could be an old eyesore that's poorly placed and annoying to try to work around.

But your thermostat is one of the most important elements of environmental comfort in your home. It controls the comfort of your home day in and day out, year round, no matter what the temperature might be outside. (And it might interest you to know that less than 50 years ago, most people didn't have central heating and airconditioning, so they didn't worry with things like thermostats.)

If you're considering automating the environmental controls in your home, your thermostat is one very important element in that automation. And it's not something that you should take lightly.

In most cases, you'll want to install some type of programmable thermostat to control the temperature of your environment and to maximize your energy savings without sacrificing comfort. One thing to remember: programmable thermostats aren't all automated. Even so, you should choose your thermostat based on your usage needs, not on the simple fact that it's automated or programmable.

There are many reasons to have a programmable or automated thermostat. The top two are the energy savings and the comfort levels that these thermostats make possible. Most programmable thermostats can help stabilize your temperature and energy usage in these ways:

1. They allow you to store and repeat multiple daily settings that help to maintain a comfortable temperature in your home without constant adjusting, which can cost you extra money when the electricity bill (or gas bill) comes in. And you can manually override these settings without affecting the rest of the program in the event that you find you temporarily need a warmer or cooler environment.
2. They allow you to store multiple temperature settings for any given day. This means that you can set different temperatures in your home based on the occupancy of your home or the activities that are taking place in your home.
3. They automatically adjust heating or cooling activation times based on changes in the outside temperature. So, for example, when it's cooler in the morning and warmer in the afternoon, a programmable thermostat will automatically adjust your environmental temperatures to compensate for those temperature changes.

Here's the fun part. Not all programmable thermostats are created equally. Some are better suited for use if you prefer to control your thermostat manually at times and others are best suited to automatic operation based on specific elements of the environment in which they are installed.

There are basically five types of programmable thermostats that range in price from around $50 to around $200, depending on the type of thermostat you decide on and the features that it has.

- **Electromechanical Thermostats**: These are usually the easiest to operate and usually have manual controls. These thermostats work with most conventional heating and cooling systems, except heat pumps. However, they have limited flexibility and can store only the same settings for each day. These are not usually used in home automation designs, as they're not as flexible as other types of programmable thermostats.
- **Digital Thermostats**: These thermostats typically have an LCD display and offer a wide range of features and flexibility. They can generally be used with most heating and cooling systems, and they provide precise temperature control. Additionally, digital thermostats usually permit custom scheduling, but they can be complicated to program.
- **Hybrid Thermostats**: These thermostats combine the technology of digital and manual controls to simplify use and maintain flexibility. Hybrid thermostats are manufactured to operate with most heating and cooling systems.

- **Occupancy Thermostats**: If heating and cooling according to the occupancy of your home is important to you, then an occupancy thermostat might be the right choice. These thermostats maintain the set temperature until someone presses a button to call for heating or cooling. The preset "comfort period" lasts from 30 minutes to 12 hours, according to how you've set the thermostat. Then, the temperature returns to the original level. These are simple, though not overly flexible thermostats.
- **Light Sensing Thermostats**: These thermostats rely on a preset lighting level to activate heating systems (they don't work for cooling systems). During lowlight conditions, a photocell inside the thermostat senses unoccupied conditions and allows the temperatures to fall 10 degrees below the occupied temperature setting. When lighting levels increase to normal, temperatures automatically adjust to comfort conditions. Light sensing thermostats are designed primarily for places where occupancy determines lighting requirements which trigger heating requirements.

The best thermostat for you will depend on your life style. Consider when you're home and how you tend to use your thermostat before you make a decision on which one to install. Then, select the thermostat that's suited to those lifestyle and usage habits.

For example, if you travel frequently and your home if often empty, an automated thermostat might be more appropriate than one that's more manual in nature. Or, if you're one of those people that wants a steady temperature but you occasionally need to temporarily change the temperature, then a digital, programmable thermostat might be more your style.

Once you've decided how your thermostat fits into your lifestyle, then you can install the one that will work best for your personality and personal comfort. And there's no better warm, fuzzy feeling than not having to worry about your thermostat, and still being comfortable.

Moving On

So, now you have a remote controlled thermostat. You'll never have to get up to adjust the temperature again. And what's even better is that this thermostat is programmable, so if you find a setting schedule that works for you, you can set it and forget it.

Home automation climate control systems take a lot of different forms. Take some time to find the one that's right for you. And then when you find it, installing it won't be very difficult. You'll be surprised at how quickly you can complete the project and how useful it will be when it's finished.

Project 5

Speak Up! Installing Voice Controls

Equipment Needed

- Windows® 98 or Higher PC
- Pentium® Processor – 266 MHz or higher
- Installed PC Sound Card
- 64 MB RAM
- 40 MB Free Disk Space for Installation; 60 MB for Normal Operation
- HAL-compatible Modem and Phone Line (required for some telephony features)
- Internet Connection (Broadband Preferred)

One of the cooler features of home automation is the ability to control some home automation by voice. It won't work with every home automation improvement that you install, but for most X-10 components it will work. And it all begins with a little program called HAL.

HAL stands for Home Automated Living, and it's basically a voice recognition program for your house. Simply put, you talk to the computer and it carries out your requests. Don't worry though. This isn't one of those programs that takes on a mind of its own and before you know it has you locked in a closet, fearing for your life.

No, HAL's gentle. And he (or she, depending on your preferences) can do nothing more than make your life a little more convenient.

For this project, we're going to install a HAL Basic system. You can find out more about the different HAL options from http://www.automatedliving.com.

Installing HAL

Installing HAL requires more than just installing some software. And there are a few little tricky parts of the installation that you should keep in mind as you walk through the installation process. For example, not all available modems support all of HAL's features. Some modems, for instance, may work with HAL's Internet feature but not with its voice mail feature. Other modems may work with the Internet and voice mail features but not the In-House Phone Interaction Feature. So, the first thing you need to do is find out if you have a HAL compatible modem, and if not, then you can purchase one from the AutomatedLiving.com web site.

Installing a HAL Capable Modem

Once you have your modem, installing it shouldn't take more than just a few minutes of your time.

1. Shut down your computer and disconnect it completely from the power, the monitor, the mouse and keyboard, and any other devices that you might have it hooked up to.
2. Then, remove the cover on the computer so that you can access the PCI Bus expansion slots. If you don't know how to remove the cover from your computer, you can probably find instructions in the manual that came with the computer.
3. Find an available PCI expansion slot and remove the slot cover if there is one in place. Save any screws that you removed when taking the slot cover off. You may need these in the future.
4. Remove the new modem from the packaging it was shipped in and insert it into the PCI

expansion slot. Firmly press down on the top edge of the modem to seat the modem into the slot.

Note

When pressing the modem into the PCI expansion slot, use equal pressure along the top so that the modem goes in straight and not at an angle. If you do not seat the modem straight, it could be damaged or may not work properly once completely installed.

5. Use the screws you took out of the PCI slot cover to secure the modem in place.
6. Replace the cover on your computer, and reconnect it to all of your peripherals. At the same time you're reconnecting your computer, you're also going to need to connect the modem you just installed to a telephone cable and then into a phone jack. HAL communicates via phone lines so you'll need both a line into the computer and a line out of it. Once you've completed the connection, turn your computer back on.

Now you have a HAL compatible modem, but that modem will probably need to have the modem drivers installed before it can be used.

Installing Modem Drivers

It's most likely that your computer won't recognize the modem, so you'll have to manually install the modem drivers for the newly added HAL-compatible modem. Since Windows XP operating systems are most common, those are the only ones we'll cover here. However, if you have a different version of the Windows operating system, you can find more information about installing the modem drivers in the HAL user's manual.

Windows XP

1. When you turn on your computer for the first time after installing the new modem, Windows will automatically detect the new hardware and will open the Found New Hardware Wizard.
2. Click Next at the bottom of the first screen. Then insert the HAL program CD that came with your HAL software into the CD-ROM drive.
3. Note: If you downloaded your HAL program from the Internet, you might not have the CD. If this is the case, you can also download the drivers from the AutomatedLiving web site (http://www.AutomatedLiving.com).
4. On the next screen of the Wizard, select **Install from a list or specific location (Advanced)** and then click next to continue.
5. On the next screen, select **Don't search. I will choose the driver to install**. And then click next to continue.
6. The Hardware Type screen will appear. Scroll down the list of devices and detect the modem you just installed and then click Next to continue.
7. The next screen is labeled Install New Modem. Select the Have Disk option and an Install From Disk dialog box appears. Click the Browse and then from the Look In drop down box, select the drive that contains the HAL CD.
8. Double click on the Drivers folder and click the small plus sign next to the folder labeled **Hal Internal PCI Voice Portal**.
9. Then double click the WindowsXP folder and select Open. The Install from Disk dialog box appears. Click OK.
10. The Found New Hardware Wizard screen appears again with **HAL HCF V.90 PCI Voice Portal VP100PCI** highlighted. Click Next.
11. At this point, you may see a Driver Compatibility Warning. Click Continue Anyway.
12. The driver for your new modem will then be installed. When the installation is finished, click Finish to close the wizard.

Now, your modem is installed and ready to go. When you begin installing the HAL software, a wizard will automatically detect the modem so you don't need to remember where it's installed at.

Installing the Hardware

Now that your computer is ready for HAL, you need to have something to connect it to once you have the software installed. Most HAL programs come with some starter hardware – either a Powerline Adapter and/or a Lamp Module. HAL can't control electrical devices like lights and appliances until a Power Line Adapter is connected to it.

Lamp modules are simple, plug in adapters, and can be quickly added to HAL any time after the initial installation is complete. The same is true with X-10 compatible devices.

Installing the Power Line Adapter

The Power Line Adapter is the equipment that adapts your powerlines to act as the conduit for HAL's signals. Therefore, the Power Line Adapter must be installed before you install the HAL software. This device has a three-prong plug on the back, a three-prong outlet on the front, and an interface connector port, on the bottom.

1. Begin by plugging the Power Line Adapter into a standard wall outlet. You should take care that the outlet you plug the adapter into is one that is always on. If you plug the adapter into an outlet that is controlled by a switch, it could be inadvertantly turned off, and then HAL will not work properly.
2. Now plug the telephone-type connector of the included Serial Port Cable and Connector into the matching outlet on the bottom of the adapter.
3. Then connect the serial end of the cable to an available serial port on your computer. This is how HAL is going to communicate between your computer and the adapters and modules throughout your house.

The outlet on the front of the Power Line Adapter is the same as a standard outlet. You can plug anything into this adapter that you would plug into a standard adapter. However, you should note that anything that's plugged into this outlet will not be controllable by HAL, because the outlet is not X-10 compatible.

Installing the Lamp Module

The lamp module also has a plug on the back of it. In addition, there is a plug on the bottom. Before you begin the installation of the lamp module, however, you need to set its address. And setting the device address can be a little confusing.

To set the module's address, which is the location that HAL will use to find and communicate with the device, use a coin or screwdriver to turn the dials on the front of the lamp module to set them to the device address. The lettered dial is the one you use to set your house code, and the numbered dial is the one to set the unit code. There are 256 possible address combinations with X-10 devices.

Now you can actually install the lamp module. The installation is simple. Plug it into the wall, and then plug the lamp that you want to control into the module.

Installing HAL

Finally, it's time to install HAL. The installation requires a little time and patience, but when you're through it, there is so much that you can do with HAL that you'll be happy you took the time to complete it.

There are two ways to install HAL. You can either install it from the CD or you can download the program and install it from the Internet. For this project, we're going to install from a CD.

Installing from a CD

First, place the installation CD into the CD-ROM drive. If the drive is configured to auto-run CDs, then an interactive demo will run. When the demo is finished, the installation menu appears. Click the Install option.

Tip

If your CD drive is not configured to auto-run CDs, then go to **Start > Run** and click Browse

(Continued)

(Continued)

(or type D:/INTALLMEU.EXE, replace D with the letter representing the drive your CD is in). Then Navigate to the drive that contains the installation CD and double click INSTALLMENU.EXE. From there you can follow the same steps used in this project.

Follow the on-screen instructions, as they guide you through installing the software. Once the installation is complete, the HAL Setup Wizard will launch automatically to help you set up your new HAL software.

The HAL Setup Wizard

The HAL Setup Wizard is designed to guide you through setting up your HAL program, which will not start until the Setup Wizard has been completed. If the Wizard fails to run automatically, or if you need to run it again you can access it by going to Start > Programs > HAL > HAL Setup Wizard.

During the Setup Wizard, shut down all programs that use COM ports and ensure that the modem is properly configured in the system Control Panel. Also remember that the Power Line Adapter should be completely installed before the Setup Wizard is run.

1. When the Wizard begins to run, the first screen you will see is the Introduction screen. Be sure to read it carefully before clicking Next to continue.
2. On the next screen, you're prompted to verify your audio capabilities. Click Verify Audio to open the HAL Audio Verification program. Follow the instructions on to verify that HAL is properly connected to your sound card and that the program can detect audio input. When you've finished, click Exit to Setup to return to the Setup Wizard.
3. The Internet Services screen appears. Select your method of connecting to the Internet and choose whether you want HAL to use that method to retrieve information from the Internet. If you select Yes and click Next you'll be taken to another Internet Services dialog box, as shown in Figure 5-1.

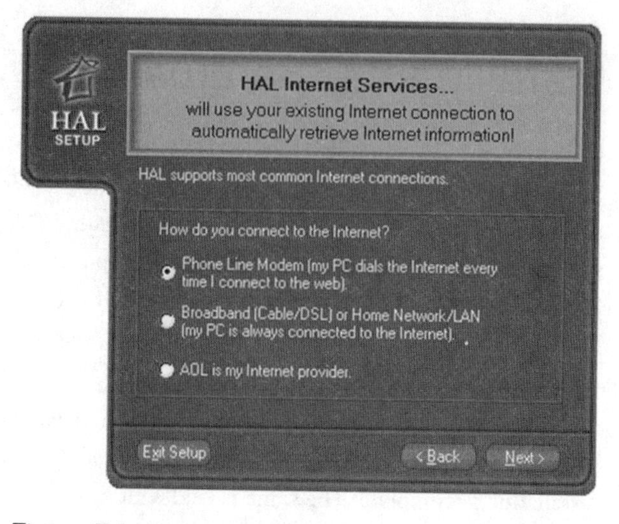

Figure 5-1 *Use the Internet Services Dialog box to set the type of internet connection you have.*

4. Select your Internet connection type, and click Next. If your internet connection is a phone line and modem, when you click next, you'll be taken to another screen where you can select the dial-up connection that you're using to connect to the Internet. After you've made your selection, click Next.
5. As Figure 5-2 shows, the next screen helps you set up weather information for your HAL system. This is an interesting feature of HAL. You can have it announce the weather forcast for the day with a single voice command. To set your weather, select the city nearest to you (or your city, if it's listed). Then click Next.

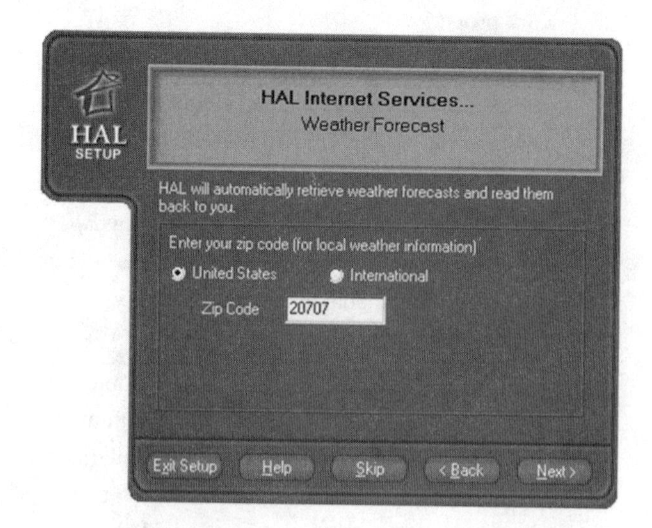

Figure 5-2 *HAL will even tell you what the weather forcast for the day is if that's what you want it to do.*

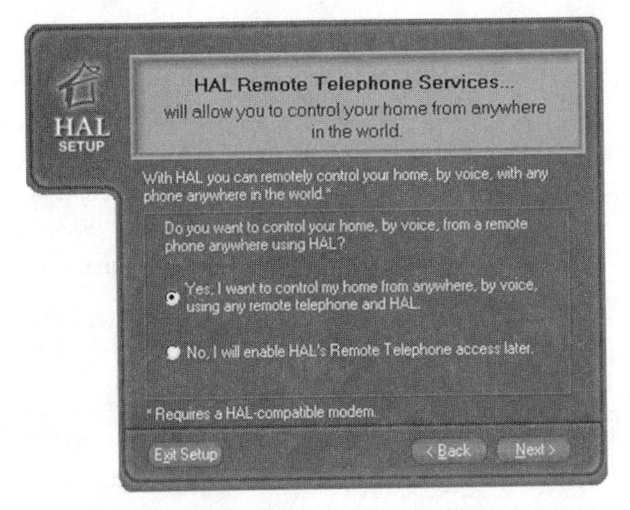

Figure 5-3 *You can communicate with HAL from an outside phone or the phone in your home.*

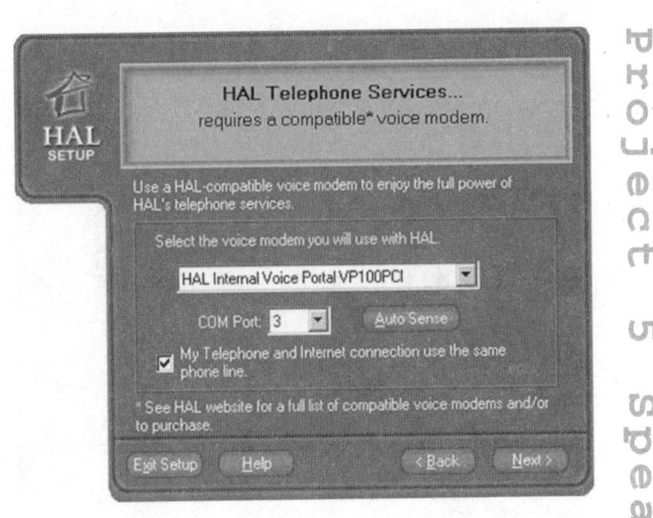

Figure 5-4 *If you're using HAL as an answering machine, you need to select your modem.*

6. Now you'll be taken to a stock portfolio configuration screen. On this screen, click Add to open a screen that will help you find stock symbols to include in your portfolio. When you find the stock symbol you're looking for, highlight it and click OK to add that stock to your portfolio. HAL can track up to 40 stocks in your portfolio at any given time.
7. When you're finished setting up your stock portfolio, click Next and you'll be taken to the screen that allows you to set up TV Listings. This option works with satellite television, and once you've selected your time zone and service provider, HAL can download your TV listing for you. You also have the option to choose a cable provider instead of a satellite provider. When you've finished making your selections, click Next to move to the next setup screen.
8. On the next screen, shown in Figure 5-3, you have the option to decide if you want to communicate with HAL using telephones outside your house. If you select Yes and click Next, another screen appears on which you should specify an access code. If you select No, you won't be able to control HAL or retrieve messages from a phone other than your home phone. When you've made your selection, click Next.
9. Next you need to decide whether you want to talk to HAL using phones that are within your home. If you select No you won't be able to interact with HAL through your home phones. When you've made your selection, click Next to go to the next screen.
10. On this screen, select your answering machine preferences. If you select Yes and click Next, you will be taken to a screen, like the one shown in Figure 5-4, from which you can select your modem. If you select No and click Next you will be taken to the E-mail setup screen. Enter the e-mail setup information requested and click Next.
11. When you complete the e-mail set up and click Next, you will be taken to a screen that lets you specify whether or not you have an X-10 Power Line Adapter. HAL lets you communicate with the electrical devices in your home via this adapter. If you select Yes and click next, you will be taken to the Hal Home Control dialog box. Make your selection, and then click Next.
12. If you do have an X-10 Power Line Adapter, select the model number on the next screen and then click Auto Sense to have HAL automatically detect the adapter. Once Hal locates the adapter, click Next to continue to the next screen.
13. Now it's time to tell HAL if you have any X-10 devices installed in your home. If you do, when you select Yes and click next you'll be taken to the HAL Home Control Services screen, shown

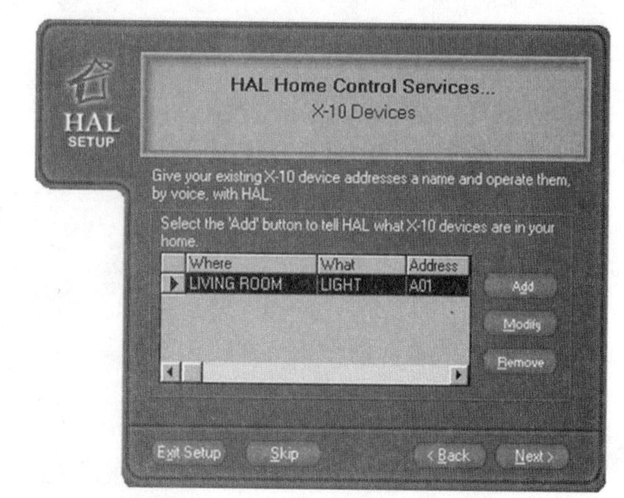

Figure 5-5 *Use the Home Control Services screen to add your X-10 devices to HAL's controls.*

in Figure 5-5. Select Add to add devices to the list that HAL will control. When you've added all of your devices, click next.

14. When you click next, the Device Wizard is activated. Walk through the Device Wizard screens to add your devices to the HAL network. You will need to give each device a name, then adjust the house and unit codes so that they match the code set on the device itself, and indicate the type of device it is. When you've finished, click Finish to add the device.
15. After you've added all of the devices that you want HAL to control, click NEXT to be taken to the final setup screen.
16. On the last setup screen, watch the introduction to how to use HAL, and then select the Run HAL Now option. Click Finish to close setup and activate HAL.
17. All that's left is to register HAL. You must register the program to use it. To register, shut the program down and then restart it. A registration screen appears.
18. Fill in the required information and click Register. The program will finish loading and you're active.
19. By default, some voice commands are already programmed into HAL. To use those commands, however, you should have a microphone connected to your computer and active. Then click the ear icon on your computer, and HAL will respond. Use one of the existing commands to tell HAL to perform a function or provide information. For example, you can click the ear icon and then say "What time is it?" and HAL will tell you the time.

> **Note**
>
> You can also enable an attention word so that when you say the word HAL becomes active. It eliminates the need to be at the computer each time you want to use HAL.

For a full list of the available voice commands, see the manual that comes with your software.

How HAL Works

When you first start considering an application like HAL, it might seem to be a home automation system in itself. But the truth is, HAL, which is short for Home Automated Living, is actually a software program that controls other home automation modules. Without those other, X-10 enabled modules, HAL is pretty much worthless.

See, HAL is software. The program is a home automation management program that manages your home automation devices. The fun part about HAL is that it's voice controlled. That means that as long as you have a microphone connected to your computer, you can simply tell HAL to "turn on the lights," or "set the alarm."

HAL features an Automation Setup window that you can use to enable HAL to completely run your household. It requires that you set the software up with the address of each X-10 module in your home and then you can quickly program the software to perform all types of home automation tasks, including: turn on and off outside lights according to natural lighting, illuminate your front walkway when a visitor approaches, and set the thermostat according to occupancy in your home.

The real benefit of HAL, however, is that its abilities are only limited by your imagination and the X-10 modules that you own. And each of these functions can be voice controlled. You can even program HAL to answer you, so that you know the program has heard and understood your commands and that they will be carried out. You'll learn more about all of the available functionality of HAL in later chapters in the book.

For now, you need to know that HAL must be able to "hear" and interpret your commands. This means that you'll need a microphone, as previously mentioned. It also means that you need to be in the same room as the microphone when you speak the commands that you want HAL to carry out. You can install a whole house microphone system, but it's a big undertaking that requires far more time and knowledge than we'll be covering in this book.

HAL's ability to answer you also requires a couple of pieces of hardware. The first, of course, is the computer. The computer will act as the translator between you and HAL. When you speak your command into the microphone, the computer will translate it into a language that HAL can understand.

In turn, HAL can answer you, if you have speakers connected to your computer, or if you have a whole house automation system connected to the computer. (There's a lot to be said for computerizing your whole home. The benefits are tremendous, but then the effort is pretty substantial, too.)

In order to answer you, HAL needs some help from the computer. When it receives a command, you can set up a condition that HAL confirms the command. For example, if you tell HAL to "turn on the lights," HAL can then respond, "would you like me to turn on the lights?" or it can respond, "of course I'll turn on the lights."

To respond in this manner, HAL needs to talk to your computer, which translates that message and then plays that response back to you. It's a complex system that requires both the ability to monitor X-10 signals and to communicate with the computer. But you don't have to worry about all of the specifics. All you really need to understand is that HAL is a computer program, and without the computer, it won't do you one bit of good, no matter how many X-10 devices you might have installed around your home.

Moving On

HAL is one of those super neat programs that lets you do many different things with little effort. What's covered in this chapter only scratches the surface of the capabilities that HAL will give you. Take some time after you install and activate the program to get to know how to use it and what you can do with it. You won't be disappointed. And the only limits you'll have are your own imagination.

Project 6

What's at the Door? Installing a Surveillance Camera

Equipment Needed

- PC with Intel Pentium 2 or Faster
- CD-ROM Drive
- 128MB of RAM
- 45MB of Hard Drive Space
- 500MB for Archives (maximum)
- USB Support

Surveillance cameras used to be only for businesses and the ultra-rich. That's not the case anymore. Today, there are excellent video surveillance systems available at very affordable prices. And as with most other types of home automation, you can start small and work up.

So, what are your options for surveillance cameras? Basically they come in two flavors: Wired and Wireless. Wired surveillance cameras require a connection to your house wiring for power and to transmit signals. Wireless cameras, however, don't require any direct wiring, can transmit signals to a wireless receiver, and are usually powered by battery.

The one real factor that you should consider when deciding between wired and wireless cameras is the placement of the cameras. Most wireless cameras cannot be located directly in the elements, and are subject to electrical problems if exposed to moisture. Wired video cameras are a little more forgiving, but can be considerably more expensive, to own and to install.

So, for this project, we're going to install a Motorola Home Monitoring Starter Kit, as shown in Figure 6-1.

The nice thing about this system is that it gives you several options, including the ability to view video clips with sound on your mobile phone or through e-mail. Additionally, this is a wireless surveillance system, so no major rewiring is required.

Figure 6-1 *The Motorola Home Monitoring Starter Kit gives you options without breaking the bank.*

Installing the Home Monitoring System

The Motorola Home Monitoring System consist of three parts: the wireless camera, the USB gateway, and the software that lets you monitor your surroundings. Installation of this system begins with mounting the camera in the desired location, then installing the software for the system, and finally connecting the USB gateway. Let's start with mounting the camera.

> **NOTE**
>
> Do not connect the USB gateway until after you've installed the system software. Connecting the USB gateway before the software is installed could cause your system to not work properly.

Mounting the Wireless Camera

When mounting the wireless camera that goes with the home monitoring system, select a location that is protected from the elements. For example, if you plan to mount the camera outside, place it where moisture will not come in direct contact with the camera. Moisture can cause the camera to short out or not work properly. Mounting it in a garage, under the patio cover, on the carport, or well behind the eaves of the house is the best option.

Another consideration during mounting is the viewing area and angle of the camera. The camera that comes with the Motorola Home Monitoring System has a viewing area of about 20 feet, and a viewing angle of about 80 degrees. So, for example, if you want to know who's at the front door, mount the camera above the door or slightly to one side and you should have a perfect view of your visitors.

When you've determined the right location for your camera, use the two screws provided to mount the camera to the wall. Depending on the location of the camera, you may need to use the drywall mounting anchors that are included, as well. To use the mounting anchors, drill a hole into the drywall using a 3/16 inch drill bit, then gently tap the drywall anchor into the hole. Once the drywall anchors are in place, you can screw the camera into the drywall mounts for better stability.

Installing the Software and USB Gateway

Once your camera is installed, then it's time to install the provided software on your computer. The software is point-and-click easy, so you shouldn't encounter too many difficulties while installing it.

1. Insert the system CD in the CD-ROM drive. The disk should auto-run and the Motorola Home Monitor Installation wizard will display.
2. If the disk doesn't auto-run, then go to Start > Run, and click browse to select the drive that the CD is in. Then, double click the Launch.exe file to start the wizard.
3. The first screen that appears prompts you to choose the Destination Location. Choose either the default location or select a different location by using the Browse button. When your selection is made, click Next.
4. Now you will be prompted to select the program folder in which the application is installed. Click Next to accept the default location or click Browse to choose a different location. When your selection is made, click Next.
5. The file installation should begin and the Install Motorola Home Monitor Hardware and Driver window is displayed.
6. Before you go any further, remove the back cover from your USB Gateway. When the cover has been removed, it should look like the gateway pictured in Figure 6-2.
7. Connect the power to the USB gateway and then connect the USB gateway to your computer using the USB cable provided. Windows should automatically install the required drivers.
8. Once the driver installation is complete, click Next on the software screen you were on before you connected the gateway. A message confirming the successful installation of the software should be displayed, as shown in Figure 6-3.
9. Check the box beside Launch Motorola Home Monitor and click Finish to launch the Motorola Setup Wizard.

Setting Up Your Software

1. Once the software is installed on your computer, you can begin the Motorola Setup Wizard. This wizard will guide you through the process of detecting and setting guidelines for your wireless surveillance cameras.

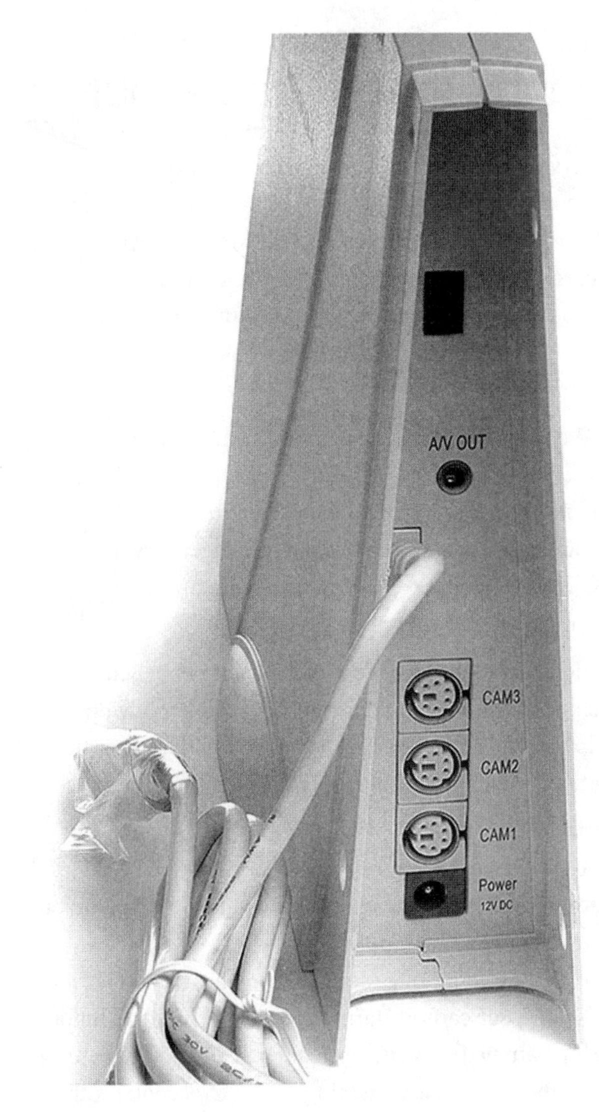

Figure 6-2 *The Motorola USB gateway allows you to connect up to three cameras.*

2. When the Setup Wizard launches, click Next. This should take you to a screen where you are prompted to detect any wired cameras that you may have connected If you have no wired cameras on your network, click Next.
3. Read through he Introduction to Detecting Devices window that is displayed and then click Next.
4. If you haven't already done it, connect the power to your wireless camera now. You'll be taken to the Device Discovery dialog box. To create a connection between the wireless camera and the USB gateway, push the discovery button

Installation of Motorola Home Monitor 2.0.01

Software installation was successful.

To continue with your Motorola Home Monitor installation please make sure that the AC adapter is plugged into the Motorola Home Monitor System Controller and that the Motorola Home Monitor System Controller is connected to the USB port on your PC. Click on the Finish button to continue

Motorola Home Monitor System Controller is connected, Launch Motorola Home Monitor.

< Back | Finish | Cancel

Figure 6-3 *The application will confirm when the software installation is complete.*

on the back of the wireless camera. (A discovery tool is provided for you to use to press the button.)

5. Once the application discovers the camera, it is listed on the Device Discovery window. Click Next.

Warning

Once the camera has been discovered the first time, you can disconnect and move the camera without needing to go through the discovery process a second time. However, the Wireless Camera must use AC power for the discovery process.

6. The next window prompts you to choose whether you have a wireless network and what type. If you have a wireless network, select the network channel from the drop down list. If you do not have a wireless network, select Don't Know. Then click Next.
7. This is the Finish Setup screen. Click Finish and your setup will be completed.

Configuring the Software

Once you have completed the installation and detection process, then you can configure your software to meet your specific needs. Start by setting up Events, which are notification and behavior guidelines for the software.

1. From the Control Panel click the Edit Properties icon for your wireless camera.
2. Give your camera a name and then adjust the settings of the camera to: 60 seconds Quiet Period Before Rearming and 10 second Video Capture Length. Do not change the brightness, contrast, or quality settings.
3. Next, click Event Setup.
4. Select your wireless camera as the sensor device and then select Motion Detected for our event.
5. Select Send to Local Archive and then click Update. Your changes will be saved, and the event setup will be completed. Now when motion is detected, a 10 second video capture will automatically take place, then the system will rearm itself after 60 seconds of no motion detection.

Now you can create a profile so that the software knows where to notify you if an event is triggered.

1. From the main menu, select Event Setup.
2. Then click Edit Profiles.
3. Select the Notification Type (in this case e-mail notification).
4. Then enter the e-mail address at which you want to be notified and click Add.
5. Once you've created a profile, it will appear in the Profiles List. Click Done to exit the Profile Setup.

You can set more than one profile for different types of notifications and switch between them as your needs change. It's also easy to change a profile by clicking the profile name.

At any time, you can also add additional cameras to the system. So, instead of only having the first camera that you installed, you can purchase separately up to two more cameras and place one on the back entrance to your house and one in your child's room if you like. Then, if you want to know if the baby's sleeping, there's no need to open the door and risk waking him or her up.

Planning Camera Layout

A surveillance system is an essential part of any automated home. The basic design of your surveillance system begins with analyzing your space and your security needs. It also helps if you have a good understanding of the hardware and software that you'll need to accomplish the surveillance design that you have in mind.

When you're considering placement for your surveillance cameras, there are two specific spaces that you need to consider. The first is your interior. That's the area inside your home where having a surveillance record of what happens (or who is coming and going) can offer after-the-fact security.

Here, after-the-fact security may also be called historical security. In your interior, it probably won't help you a lot in an emergency situation. But after the fact, the video record of the events that took place during the emergency can be invaluable.

Your second monitoring consideration will be your exterior, and it's in the exterior area that you'll find you have some benefit from having surveillance cameras available to you. For example, if you hear a strange noise on your carport during the middle of the night, a well positioned surveillance camera can allow you to see if the noise you're hearing is just a raccoon rooting around in the trashcans or someone trying to break into your house, car, or shed.

The key here is that your cameras be installed in the best possible location. And when you're considering installation, depending on how you want to view your environment, you may need either wireless or wired cameras.

- **Wireless Surveillance Cameras**: These cameras use battery-powered radio transmitters and receivers to connect to your monitoring station. The basic advantages of wireless security systems are that they are easy to install (you're not confined by wiring) and they're relatively easy to set up and use. The very nature of wireless surveillance cameras also allows you to set them up in areas where you may not be able to install wired cameras.

- **Hard-wired Surveillance Cameras**: Hard-wired surveillance cameras are operated by connecting the wiring installed inside the walls, attics, crawl spaces, of your home. In some cases, hard-wired cameras may also be connected to underground wiring if that's the type of wiring that's prevalent in your home. Hard-wired systems are usually installed by a professional contractor and can be less visible and more aesthetically pleasing than wireless components.

Whether you use wireless or wired cameras, or some combination of them will be determined by your specific needs. And that's where the layout of your cameras need to be considered.

When you're installing surveillance cameras for security purposes, you'll want to install enough cameras so that you have some overlapping views of the areas that you want to monitor. For example, if you're monitoring the perimeter of your house with surveillance cameras, you'll probably want to put two cameras at each corner of the house so that you can view the perimeter of the house completely, and from different directions.

In this manner, you'll get the most complete picture of the perimeter of your home. It's the same for inside your home as well. If you want to monitor your front door, you might have cameras set in two different places, and even at two different angles watching that front door. That way you'll not only get a good picture of who's coming through the door, but also where they're headed once they're inside the house.

Another aspect to consider about the placement of your surveillance cameras is the angle at which the camera records. This angle is cone shaped, with the tightest part of the cone being located at the closest point to the camera and expanding outward, and it differs some from manufacturer to manufacturer. Keep these measurements in mind as you're placing your cameras, and don't hesitate to readjust them if they're not capturing the areas that you're most interested in monitoring.

One last consideration when installing surveillance cameras is the distance at which they will record movement or activity. Surveillance cameras often don't offer the same type of picture that you might get from your video camera. Even if they did, at a certain distance from the camera, the picture is no longer clear enough to discern more than basic movements. The same will hold true for your surveillance cameras.

Be sure that you're familiar with the optimal recording distance for your security camera and then set it up so that the area that you're most interested in recording falls within that distance. Otherwise, what you could end up with are pictures that just aren't clear enough to define what's happening, exactly.

There's more than just mounting your surveillance cameras if you want them to be effective. Take the time to position them properly to ensure that the records that result actually show the results that you mean for them to show.

Moving On

Having the ability to monitor the entrances to your home or to monitor rooms within your home will give you a feeling of safety and security. Installing a wireless surveillance camera is the way to achieve that ability. It's easy to do, and it takes only an hour or so to complete. And when the initial installation is done, it's easy to add additional cameras or monitoring equipment to the system.

Project 7

Shhh...I'm on the Phone

Equipment Needed

- HAL2000
- PC

Don't you hate it when you're on the phone and someone insists on talking to you? Most people do, and it's even worse when you're on the phone and the house insists on talking to you. You can't follow either what the house is telling you or follow the telephone conversation that you're having.

The HAL program that we've discussed in previous chapters has some neat telephone capabilities, including the ability to shut up when the phone rings. This lets you have the conversations that you need to have without the hassle of trying to talk over the house.

In this project, we're going to learn about configuring the HAL telephone system so that you not only can have your conversations in peace, but also so that HAL will help you out with some of the incoming calls that your receive.

Caller ID Configuration

Caller ID is one of the most useful telephone tools ever invented. When someone calls, their number is registered on a small device that tells you what the number is and who it belongs to. And now, HAL can tell you who is calling. All you have to do is set up the Caller ID screen.

1. To get to the Caller ID screen, right click on the ear icon and then select Open System Settings. The Telephone screen will appear.
2. On the Telephone screen, click the Caller ID option. You'll be taken to the Caller ID screen shown in Figure 7-1.

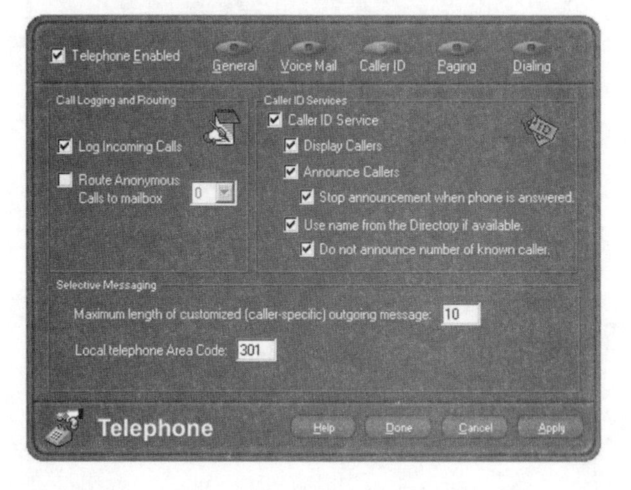

Figure 7-1 *The Caller ID screen is where you set up your caller ID options.*

3. On the Caller ID screen, you can set your options for how anonymous calls are handled, what caller ID capabilities are available, and selecting messaging options. Select those options that work for your specific situation.
4. When you're finished, click Apply to apply the settings and remain in the menu. If you want to get out of the menu, you can click Done and your settings will be saved and the screen will close.

Configuring Voicemail

1. You can also configure your voicemail from the Telephone screen. To configure voicemail, select Voicemail from the top menu. You'll be taken to the Voicemail screen, shown in Figure 7-2.
2. On the Voicemail screen, you can choose voicemail settings and message options.
3. You'll also find a control that lets you configure when HAL will answer the phone if you're not available. Well, technically HAL doesn't answer the phone, he sends it to voicemail, but you can tell him when to do that from this screen.

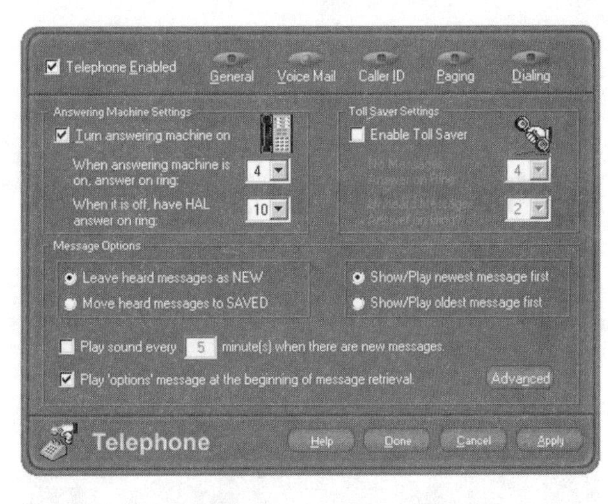

Figure 7-2 *Use the Voicemail screen to set your voicemail options.*

Figure 7-3 *All of your Internet-related HAL options are on this screen.*

4. When you're finished making your telephone configurations, click Done. Your settings will be saved and you will be returned to your desktop.

Configuring Other Settings

HAL also has some additional features that you might find useful. For example, did you know that you can get HAL to tell you what the weather will be for any given day or what the stocks and top headlines for the day are?

You can. But you don't access them from the Telephone page that you've been working on. Instead, these are located on the View Internet Information screen. You can get to it by right clicking the ear icon.

When you do, you'll see a screen similar to the one shown in Figure 7-3. This screen lists all of your Internet options. As long as HAL is connected to the Internet, you can use these options however they suit you.

Home Entertainment Servers

You may have heard them called Home Media Servers, or Home Entertainment Servers, and even Home Television Servers, but these devices perform a very neat function – they connect to your television and make it available via a network.

Why would you want your home entertainment devices, including your television, available via a network? Well, for starters, that network would give you access to your entertainment media no matter where you happened to be, whether it's another room in your home or from a location completely outside your home.

Networking your home entertainment equipment also makes management of that equipment easier to manage. Basically what happens is that you connect your television to the home entertainment server and then connect the home entertainment server to your network, and the next thing you know, you're connected all the way through the house.

Of course, that's a pretty simplistic view of the process. If you want to see how truly difficult it is, you can check out this sample chapter from another home automation book at http://www.extremetech.com/article2/0,1558,1813430,00.asp. That article details (very well, I might add) exactly how you would go about creating a home entertainment server.

You don't have to create your own, though. There are a few companies, such as Pioneer and Toshiba that make devices that work in a very similar manner that are easier to use and don't require hours of component building to create them. These devices vary in price from about $300 up.

Once you have a home entertainment server, and with some additional equipment, you can connect all of the televisions and other home entertainment devices in your house so that you can access them from any room without the need to purchase or lease additional devices from the cable or satellite company. The rental on these devices can run you as much as $25 per week, depending on where you purchase them from.

When you get the device integrated into your network, you'll have the freedom that you can only dream about without a server. For example, you can connect all of your televisions to the home entertainment server without the need to add an additional cable or satellite box at an additional monthly fee.

The one downfall with a home entertainment server is that they are not yet designed to be used wirelessly, therefore, you'll have to run cable from the server to each device that you want connected to the device.

It's a bulky, clunky way to have a home entertainment server, but until a wireless alternative becomes available (and it shouldn't be long now) that's the best option that you have if you feel the need to add a home entertainment server to the mix.

Moving On

HAL has some easy to use interfaces that help you get as much out of the system as you possibly can. From answering the phone to providing you with the top news stories, just because you asked, the program is the next best thing to a personal assistant.

Of course, HAL can't do everything. But even so, the program is one of the must useful home automation software applications you'll find.

Project 8

Music Everywhere: Audio Controls

Equipment Needed

- PC
- Hi-Fi Link from Xitel
- HAL Digital Music Center Software

So, you want to get control of your whole house audio? No problem. HAL (remember him from back in Project 5) can give you all the control you need. For example, maybe you're in the kitchen cooking. Your hands are all sticky and you don't want to wash them just to turn the stereo on. No problem. A simple phrase like, "Hal, turn the music on in the kitchen," will have you dancing on the counters. And it's not nearly as difficult as it sounds.

There is one catch, however. This project assumes that you have a wired whole-house audio system set up. If you don't, no worries. You can use this project to give HAL control over your central stereo system, too. And if you live in a small place, that central stereo system just might be your whole-house audio!

Connecting the Hi-Fi Link

To start this project, you need a way to connect your stereo system so that HAL can control it. Easy enough. If your computer is within 30 feet of your stereo system, you can connect the two, and give HAL control. But you need the Hi-Fi Link from Xitel to do it.

The Hi-Fi link is a simple system that requires no software installation.

1. Unpack the contents of the Hi-Fi Link package.
2. Connect the RCA jacks to your stereo and then connect the opposite end to the Hi-Fi Link.
3. Now, connect the USB cable to your PC and then connect it to the USB port on the Hi-Fi Link. Your computer should automatically detect and install the appropriate drivers for the Hi-Fi Link.

Once you've set the Hi-Fi Link up, then you can add it as a device that HAL controls and install the HAL Digital Music Center (DMC), which you can use to control your music using nothing but your voice.

Note

The HAL Digital Music Center is an add-on software module. You cannot use this product without the HAL Home Control software installed and operational. If you do not have this software installed, please go back to Project 5 and follow the instructions there for installation.

Installing the Digital Music Center

HAL's Digital Music Center is actually a voice controlled music streaming application. What that means to you is that HAL will use the music that you have stored on your computer, and using the Hi-Fi Link you installed in the last step, you can play that music through your stereo speakers. Installing HAL's DMC gives you voice control over your music, and it even has the capability to rip music from any CD that you place in your CD drive. So, you can create and change play lists from your music or add new music to your collection from any CD that you own or purchase.

HAL DMC is a download-only program, so once you download it from the AutomatedLiving.com web site, to install it all you need to do is unzip the file and double click the installation icon. The HAL DMC program should install automatically.

Setting Up DMC

When the installation is complete, you need to run the DMC Setup Wizard. Right click the ear icon on your desktop and select Open Digital Music Center.

1. You'll be taken to the main DMC screen, shown in Figure 8-1. The first time you open the program, the screen may include some introductory information about the DMC program. Read through this information and then click Next.
2. On the next screen you need to tell HAL where to scan for music on your computer. DMC will detect MP3 and Windows Media files when you tell it where to look for them. To do that click Add, and then navigate to the location where you store music files.
3. Double click the music folder to select it. Alternatively, you can click the option, Scan All Drives, and DMC will automatically scan all of the drives connected to your computer. When you've made your selection, click Next.

Figure 8-1 *The main DMC screen is your portal into the program.*

4. The next screen that appears is the progress meter for the scan. When the scan is complete, you'll see a message that indicates the scan is finished. Once you see this message, click Next to continue.
5. DMC will show you how many music files have been located. Click next to continue.
6. Now DMC is ready to look for music that you have recorded from your personal CD collection. Select the storage location for your recorded music.
7. At this point, you also have an option to have the program connect to the Internet to collect information about the songs you have recorded. The information is used to categorize each song by title, artist, album, year, and genre. If you want DMC to perform this search, select the option and then click Next.

> **Note**
>
> You must have HAL connected to the Internet for DMC to be able to search for song information.

8. Finally, you'll be taken to a screen that includes information about the DMC. Read this information and then click Finish to be taken to the DMC main screen.

Adjusting DMC Settings

Once you get to the main DMC screen, you should adjust some of the settings so the program operates properly.

The first setting you'll want to adjust is the Player Confirmation settings. To do this, click the settings button on the DMC screen, shown in Figure 8-2. You'll be taken to the DMC Settings screen.

On this screen, shown in Figure 8-3, you can select whether you want to enable or disable the Command Confirmations option. A command confirmation is when HAL confirms the command you have just given it. For example, if you want the music to start, you can say "Play music in the kitchen." HAL will confirm the

Figure 8-2 *Click the settings button to change the Player Confirmation Settings.*

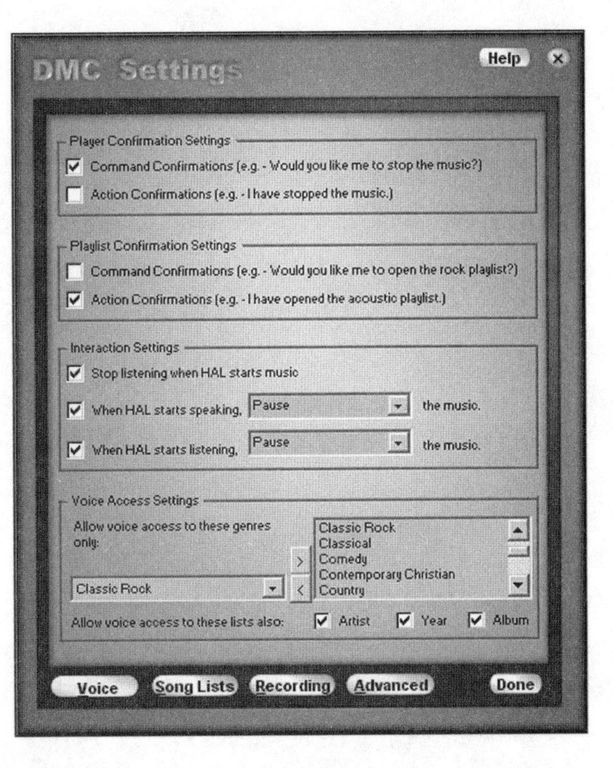

Figure 8-3 *Use the DMC Settings screen to change your confirmations, interactions, and voice access settings.*

command by asking, "Would you like me to play music in the kitchen." To enable this option, place a checkmark in the box next to Command Confirmation. If you want to disable the command, remove the checkmark from the box.

You'll also find an option to enable or disable Action Confirmations on this page. If Action Confirmations is enabled, then when HAL turns on the music in the kitchen, it will say, "I have turned on the music in the kitchen." To enable this option, please place a checkmark in the box next to enable. To disable, remove the checkmark.

There are also Command Confirmations and Action Confirmations for your play list settings. These confirmations work in the same way described above, but they apply specifically to your play list selections.

Next you'll find Interaction Settings. It is on this screen that you'll find the command Stop listening when HAL starts music. If you select this option, then HAL leaves the listening mode when the music starts. That means that you cannot ask HAL to perform any functions while the music is playing.

You can also choose from a drop down menu the action that you want HAL to take with the music when HAL is about to speak. For example, when an event (such as the phone ringing) creates the need for HAL to speak to you, you have the choice to Stop, Mute, or Pause the music so HAL can speak. The same options are available when HAL starts listening.

Finally, you'll find Voice Access settings near the bottom of the screen. You can use these settings to select the genres you want to be able to play by voice. So, if you want to request that HAL play a specific type of song, as long as that genre is on the list to the right, you can make that request. To add a genre to this field, highlight it on the drop down menu and then click the right arrow key. To remove one, highlight on the list at right and then click the left arrow key. You can also enable voice commands for the Artist, Year, and Album by placing a checkmark in the box next to each option.

One last option on the DMC Settings Screen that you should be aware of is the Advanced option. To view your advanced settings, click the Advanced button on the bottom of the screen.

The Advanced Options screen, shown in Figure 8-4, is where you can tell HAL where you want to hear your music played. To add a location to your list, type the name of the location (i.e. Kitchen), and then click the right arrow to add to the location to your list. If you

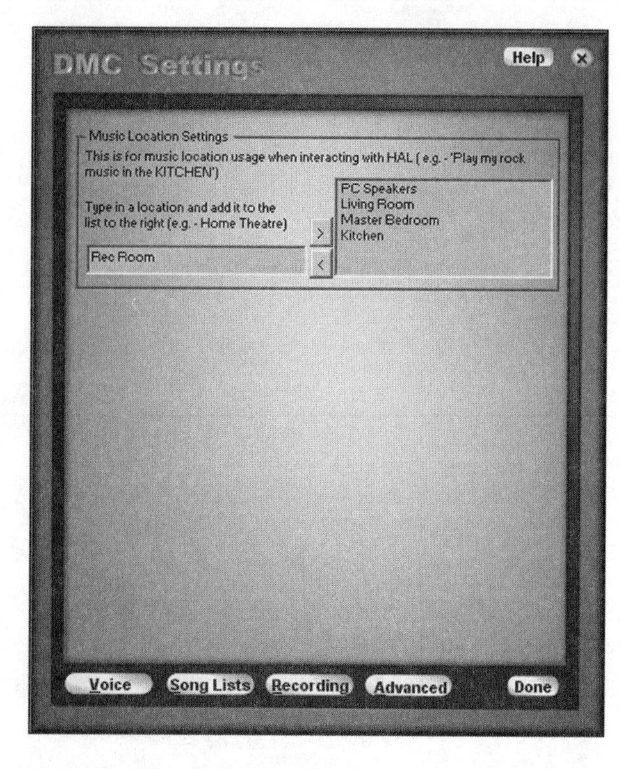

Figure 8-4 *Use the Advanced Options screen to tell HAL what rooms you have set up for music playback.*

have speakers throughout your home that are controlled by an automation controller, then you can use this feature to play your music only in the requested room.

Creating Song Lists

Once you have the DMC installed and configured, then you'll need to create a song list before HAL can play music for you. You can set up a song list from the main DMC page by clicking the Add option in the Current Song List box on the right side of the page.

In the dialog box that appears, type a name for the song list that you're creating and click save. Then all you need to do is drag and drop songs from the Master Song List to the Current Song List field.

When you've added all of the songs that you want to appear on the song list, click Save and your new song list will be created. You can create numerous song lists using this method. Then when you want to play one of your song lists, simply ask HAL to play that song list.

Choosing an Automation Installer

There may be some home automation projects that you just aren't sure you can accomplish. For example, some of the projects that we've done so far are easy projects that amount to plug-and-play home automation. On the other hand, there are some projects, especially when you get into audio and video control systems, which require a lot of specialized knowledge. And at some point you may well decide that you need to bring in a contractor to handle the hard stuff.

If you have decided that you need some help with one (or all) of your home automation projects, knowing which contractor to hire can be almost as frustrating as the whole home automation process itself. Fortunately, there are a few resources that you can turn to online if you've decided that you want to hire someone to help you out.

- The X-10.com web site contains a searchable database of home automation installers. Enter your zip code to find the registered installers in your area. (http://www.x10.com/pro/installer_home.htm)
- Smarthome.com also has an installer referral program. Enter your information into the form provided and click **Submit** to be taken to a list of installers in your area. You can then finish the referral process to have one of the contractors call you to arrange an appointment to discuss your needs. (http://www.smarthome.com/asp/projref.asp)
- Home-Automation.org has a list of home automation installers that are arranged alphabetically by location. There's no special buttons to click or forms to fill out on this site. Just select the installer that interests you, and you'll be taken to that company's web site. (http://www.smarthome.com/asp/projref.asp)

Of course, picking an installer from a list or referral often isn't enough to ensure that you're hiring a contractor that will do quality work. You should take the time to interview several contracts before you give the job to one of them. Also have each contractor quote you a price, but remember, price isn't everything when it comes to this type of construction work.

When you interview a contractor, you may want to ask them some very specific questions about their experiences with home automation projects. For example, here's a short list of questions that you should ask your potential contractors:

- How much experience do you have with X-10 home automation?
- How long have you been working with home automation projects and how many have you completed?
- Who were some of your previous clients and may I contact several of them?
- How much will your pricing vary?
- Are there hidden costs that I should be aware of?

When you contractor answers these questions, there are some clues that you should watch for to help you determine if a contractor is legitimate and knows enough about home automation projects to successfully complete your project. For example, if you ask the installer how many projects he or she has completed and they hedge on the answer, it could indicate that the installer hasn't actually done home automation projects in the past, but they are interested in it.

Also be wary of any home automation installer that refuses to provide references or provides references that are older than about a year old. Working contractors will always have current references that they are happy to have you contact. If not, don't even consider the contractor.

And make it a point to call **all** of the references that your potential contractors provide. It's not a fun chore, but many installers will provide clients lists that are bogus, banking on the fact that most people do not call the references which they request.

When you do call a reference, be sure to ask them what type of project the installer performed for them, how the project went where both time and money are concerned, and if the customer would ask the installer back to do another project. If the reference on the other end of the line is real, they will likely be willing to answer all of your questions very candidly. After three or four calls, you'll have a good feel for the type of contractor that you're considering.

There are also two more considerations that you may want to keep in mind as you're making the decision on what contractor to hire. The first consideration is the type of guarantee the installer offers on the work they're doing. When you're working on wiring and electronics, any number of problems can occur in the first few months after the installation. If the installer that you hire offers a guarantee on their work for a specific period of time after the installation, this will help to ensure that you'll remain happy with your new home automation capabilities.

Finally, be sure that you get a written quote from each installer that you're considering working with. Too often, unwritten quotes will change (and usually become much higher) after the deal has been made. A written, and signed, quote from the installer will help protect your pocketbook in the long run.

There's no shame in admitting that you can't, or simply don't want to, finish a home automation project. Some of them are very complex and require more technical skills that even the best do-it-yourselfer might have. When you find that a project is just more than you can accomplish, don't hesitate to call in help. Professional installers will make your life easier and your automation projects far more successful.

Moving On

That's it. Now HAL is set up to stream your music through your house...or at the very least into the rooms in which you have stereo speakers installed. You can even create your own custom music lists with only your favorite songs, or with songs from a single album. How you arrange it is entirely up to you. And because your music is now connected to HAL, you'll have far more control over how it's grouped and how you want it played.

You can even set HAL up to play music at a certain time of day, or when a certain event takes place in your home. For example, when you walk in from work, you can have HAL automatically turn on certain lights and start playing your favorite Jazz music. But that's a project you'll learn more about in Project 18.

Project 9

Is Anyone There? Installing an Occupancy Sensor

Equipment Needed

- Wireless Occupancy Sensory
- Wireless X-10 Compatible Receiver
- HAL

Occupancy sensors (which are just a fancy name for motion detectors) are useful tools for doing some of the neatest things with home automation. For example, an occupancy sensor is at the heart of a series of events – called a scene – that take place when a room becomes occupied.

For example, say you have an occupancy sensor installed in your foyer and that occupancy sensor is connected to a home automation management application like HAL. You can set up a scene that will automatically turn on lights in strategic places through the house, turn the heat up or the AC down, and turn the television on and tune it to a specific channel. All because you walked through the door. You don't have to walk around switching on lights, adjusting the climate control system, and turning on the television. HAL will do it all for you, but only because an occupancy sensor was activated.

For this project, you'll use a wireless occupancy sensor, a wireless receiver, and the HAL program. The wireless occupancy sensor is quick and easy to install. This is not the only motion sensor available, however. There are several different types of occupancy sensors, including those with light switches that will mount nicely into existing openings in the wall.

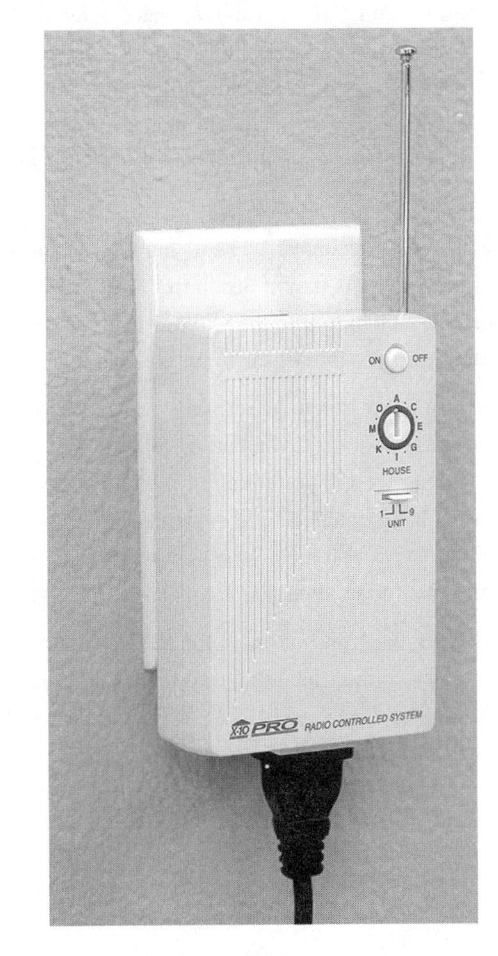

Figure 9-1 *A wireless receiver will convert wireless signals into X-10 signals that can then be controlled by programs like HAL.*

Installing the Wireless Receiver

Wireless receivers like the one shown in Figure 9-1, convert wireless signals to X-10-capable signals that you can then use with a HAL-type automation management program.

Most of these devices are simple, plug-in-play devices. So, all that's required to install this device is to plug it into a wall outlet. Then, set the house code, and you're ready to begin using it to convert signals from your wireless motion sensor.

Installing the Wireless Occupancy Sensor

Wireless occupancy sensors are designed to detect motion and send a signal to a receiver which then communicates via X-10 signals to your management application. The wireless occupancy sensor by Version II, shown in Figure 9-2, is also capable of detecting light and dark, so it can even be used to trigger an automatic event based on the available light. This is an excellent feature if you want a certain set of events or actions to take place when the light in a specific area changes dramatically.

Most wireless occupancy sensors are easy to install. All you need to do is mount the sensor either by placing it on a shelf, attaching it to the wall, or using a Velcro attachment to place the sensor in the desired location.

Once you've installed the sensor, you need to set the sensor's house code so that it can be read by the X-10 enabled receiver you installed in the last step. In most cases, X-10 enabled devices default to an A-1 house code, but if you have your system set to a different house code, then you can change it.

1. Locate the Unit button on the device (under the battery compartment lid) and press and hold that button until the red LED light flashes. The light will flash a number of times. Count these numbers; that's the number of the current house code.
2. To change the house code, press and release the Unit button the number of times that you want to set your house code to. So, for example, if you want to set your house code to "D", then press and release the Unit button four times.
3. The last time you press the button, hold it for a few seconds. The red LED light should flash to confirm your setting. It will then flash the number of times you set for the house code.
4. Use the same procedure to change the Unit Code.

Depending on the model of wireless occupancy sensor that you've selected, you may also have options to set the sensor only to changes in the light in a room, and you may have other options. If so, use the manual provided by the equipment manufacturer to complete those settings.

Once your wireless occupancy sensor is set up, then you can add the device to your management program and create scenes that will be triggered by motion or changes in the light level.

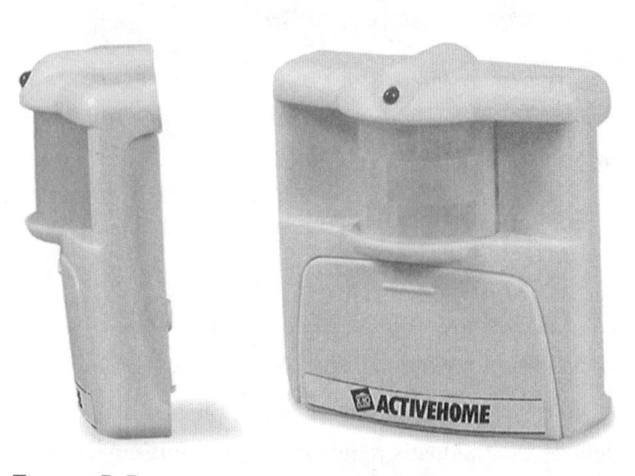

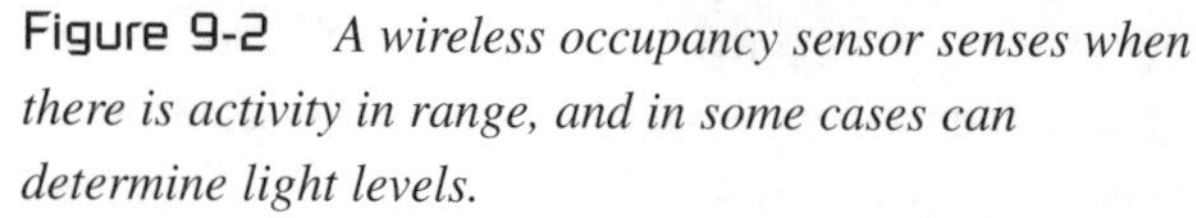

Figure 9-2 *A wireless occupancy sensor senses when there is activity in range, and in some cases can determine light levels.*

Setting the Scene

So, what can you do with an occupancy sensor? For starters, you can use it to trigger a room scene. A room scene is a group of actions that happen, based on a single event, like the activation of a motion sensor.

HAL AUTOMATION Devices Modes Schedules Tasks Sensors Directory Help

This screen allows you to manage the Devices configured for your home. It can also be used to manually control and test Devices.

Where	What	Dim	Address	Type	Verbal Confirm	Action Confirm
BACK YARD	LIGHTS	X	B01		X	X
COFFEE	POT	X	K02		X	X
FAMILY ROOM	DRAPES	X	F13		X	X
FRONT YARD	LIGHTS	X	F10		X	X
HALLWAY	LIGHTS	X	H07		X	X
KITCHEN	LIGHTS	X	K01		X	X
LIVING ROOM	LIGHT	X	A01		X	X
MASTER BATH	LIGHTS	X	M14		X	X
MASTER BEDROOM	LIGHTS	X	M05		X	X
MASTER BEDROOM	SHADES	X	M02		X	X

Report Devices Add Modify Remove Done

Figure 9-3 *Select the sensor button to add an occupancy sensor.*

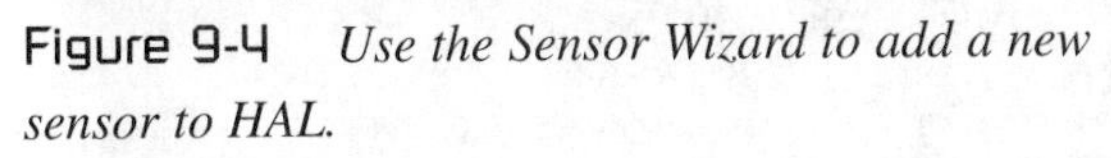

Figure 9-4 *Use the Sensor Wizard to add a new sensor to HAL.*

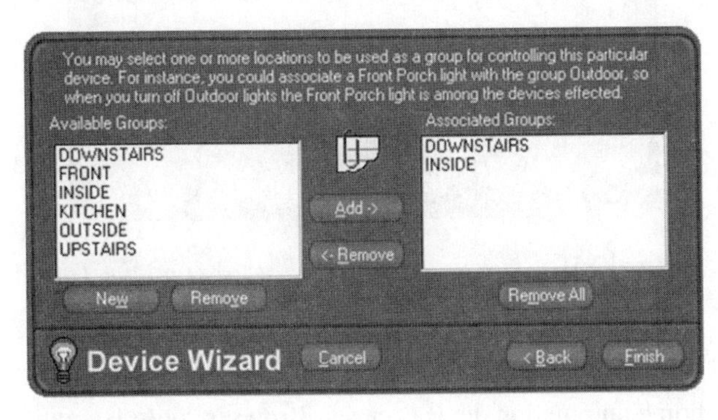

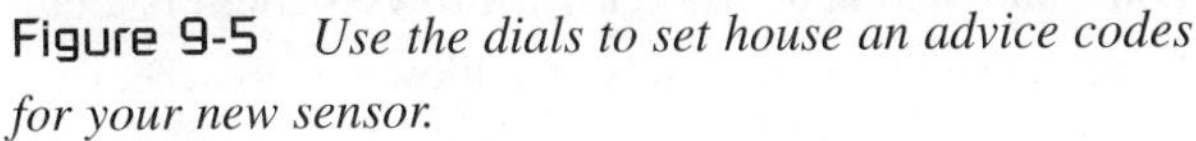

Figure 9-5 *Use the dials to set house an advice codes for your new sensor.*

For example, if you set an occupancy sensor up in your living room, then you can set it to trigger a scene so that when the sensor is activated, the lights in the room all come on at 30 percent brightness, the television comes on and automatically tunes to a specific channel, and the temperature controls for the room are adjusted to 76 degrees or whatever temperature you find most comfortable.

All it takes is a little configuration on your part. Of course, all of the devices that you want to control should be set up with X-10 control modules.

Add the Occupancy Sensor

1. To begin setting up your scene, you must first add the occupancy sensor to HAL's sensor list. To do that, right click the ear icon and select Open Automation Setup Screen. Then click the Sensors button at the top of the screen shown in Figure 9-3.
2. Click Add to add the sensor and the Sensor Wizard, shown in Figure 9-4, is displayed.
3. Give your sensor a name and then select the type of sensor it is from the Type drop down list. In this case, we're using an X-10 sensor. When you've made your selections, click Next.
4. Using the dials on the next screen (shown in Figure 9-5). Set the address for the sensor and then click Finish. Now HAL recognizes your sensor.

Create an If/Then Situation

Now that you've added your sensor to the HAL program, it's time to set up a scene that's triggered by the sensor. To do that, right-click on the ear icon and select Open Automation Setup Screen. When the screen opens, click the Tasks button near the top of the screen.

Before you go on, you need to understand how the If/Then scenario works. If is the rule that sets off a group of actions, and Then is the actions that will be activated. So for example, you can set up an If/Then situation like this: If the sensor is activated, Then the lights in the living room come on.

1. To add a rule – also called a condition – click the Add Condition button on the left side of the screen.
2. Type a name for the rule and then click OK. You'll be taken to the conditions screen, shown in Figure 9-7. Select Trigger Event or Secondary Event. A Trigger event is the primary event that you want to trigger a circumstance. The Secondary event is an event that must also be met once the trigger event is activated. In this case, select Trigger event and then click Next.
3. On the next screen, you need to select the trigger event that needs to take place for the Then scenario to be carried out. Select Sensor and click Next.
4. On the next screen, select the sensor you're setting up from the list, choose whether the event should occur if the sensor is triggered from the on or off setting, and then click OK.
5. Now you need to add the action portion of the rule. To do that, go back to the Task screen and click Add Action.

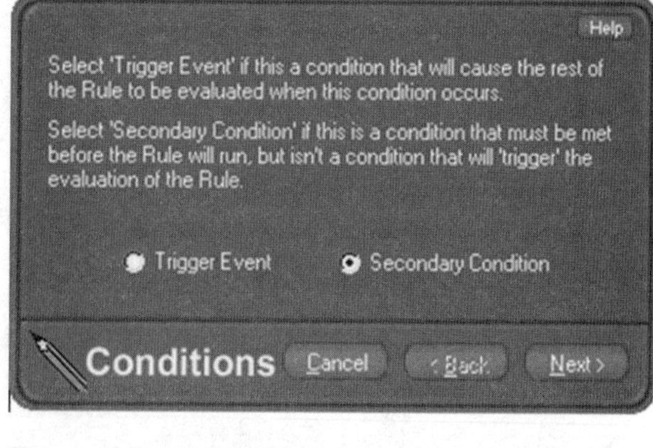

Figure 9-7 *Select Trigger Event or Secondary Event to begin the event setup.*

6. The Action Wizard, shown in Figurer 9-8 appears.
7. In the Action drop down menu, select Run Room Scene, and the available room scenes will be displayed.
8. Select the Room Scene you would like to have run when the sensor is triggered.

Note

If no room scenes are available, you can set one up by going to the Modes screen and clicking Add Scene.

9. Then select whether you want a delay in time from the trigger of the sensor to when the scene runs, and choose to confirm or not confirm the action.
10. When you've made your selections, click OK and your If/Then scenario will be complete. Now, when the sensor you've installed is triggered, the desired actions will take place.

Choosing the Right Browser Controls

When you're putting together multiple home automation projects, you may also want to add browser controls for those projects to make managing your

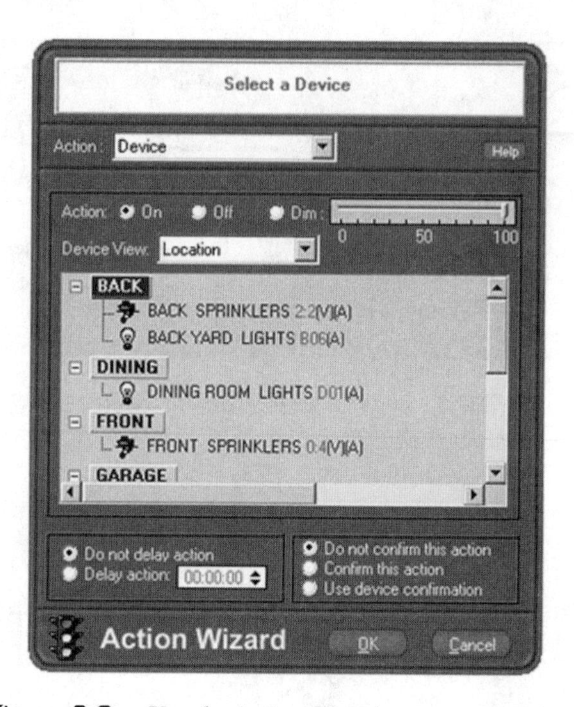

Figure 9-8 *Use the Action Wizard to set up the Then scenario.*

home automation devices easier. Browser controls are basically software programs that allow you to access and control your home automation devices via the Internet.

HAL, which we've already discussed, is one type of home automation management system. And as you've learned, HAL is one of the very few – and perhaps the only one – that's voice controlled. HAL even allows you to connect to the software via the Internet to control your devices while you're away from your home.

But HAL isn't the only software on the market. There are several other types of home automation software programs that you can choose from as well. Here's a quick list of some of the most popular home automation software management programs on the market:

- **MisterHouse**: (http://misterhouse.source forge.net/) MisterHouse is an open source home automation management software. What this means is that the source code for MisterHouse is available to anyone that wants to change or improve it. Now, if you're not a computer programmer you might not care about having the ability to change and improve the management software that you're using. But there's another reason to use MisterHouse to control your home automation, too. Because it's open source software, it's free. All you have to do is download it, and the beauty of the program is that it will run on both Windows platforms and Unix-based platforms (such as Mac and Linux). It's not an easy to use program, but if that's your thing, then this might be the software that you're looking for.

- **HomeSeer**: (http://www.homeseer.com/) HomeSeer is a control and management software that allows you to connect different brands of X-10-enabled devices and control them from a single remote or location on the web. This is a highly recommended management program, because it's as easy to use as can be expected of any home automation project.
- **HeyU**: (http://heyu.tanj.com/) HeyU is a very basic, easy to use home automation control software. The software isn't one of the most popular on the market, but it is text-based so it's not difficult to implement, and it's also an open-source project, so again, it's a free program that can increase the value of your home automation projects (both to you and future owners of your home).
- **Xtend**: (http://www.jabberwocky.com/software/xtend/) This Unix-based home automation software is a little more advanced than some of the others that have been mentioned above, but it's also a little more functional. It's also compatible with HeyU, for a more complete application. And like HeyU, this is an open source project, so it's free.

The thing that you have to keep in mind when looking at home automation management software is how you'll use it and the platform on which you'll be installing it. Not all home automation management software is cross-platform. And not all home automation software will fit nicely with the way that you're most comfortable using it.

Take your time when evaluating home automation management software and make sure that it will perform all of the functions that you need for it to perform without needing too much tweaking and without you having to learn an entire programming language to control it.

Since most of the programs mentioned above or in other places in this book are free, you should be able to evaluate them without too many difficulties or frustrations. There is one thing that you want to keep in mind when installing open source software however.

Like many Windows-based application, it can be a little buggy, and it's not usually monitored by a formal management body, so bugs could take some time to fix. Usually it's the users that do the fixing, though, so once it's fixed, you can be pretty sure that it will stay fixed, even if it does change as users add more, and more useful components to it.

Moving On

Now you're done. Your motion sensor is installed, and you've added it to HAL and created an If/Then scene that will trigger a group of actions (or a single action if you prefer) when the sensor is activated. You can change the scenario at any time, or add to it if you prefer.

Next we're going to learn how to automate a reminder with HAL, so keep reading.

Today Is Trash Day: Automating Reminders

Equipment Needed

- Windows-based PC
- HAL2000

One problem that many of us have is remembering appointments. Even recurrent events, like taking the trash out to the street one day a week, can get away from us. Life is just so busy and there are far more important things to remember (like that doctor's appointment you have tomorrow). So, wouldn't it be nice if you could have your house remind you when you have an appointment or when you need to take the trash to the road?

You can. The HAL2000 system is a more advanced version of the HAL system that we've been looking at up to this point in this book. One of the new features of HAL2000 is a reminder service. You tell HAL when you have an appointment, and it will remind you at the appropriate time.

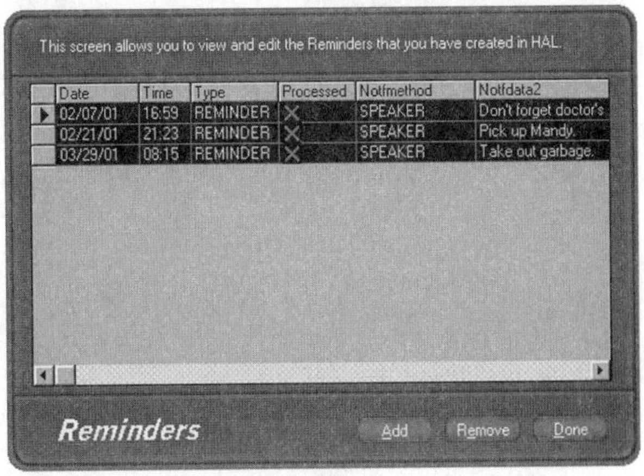

Setting Up Reminders

So, how do you get HAL to remind you of appointments or tasks that you need to complete? It's easy. But before we get too deep into it, there are few changes in the HAL2000 system that you should be aware of.

For starters, the ear icon that you used to create events in HAL in previous chapters won't be used in this chapter. Instead, there is a phone icon. That's the icon that we're going to use.

Additionally, this version of HAL has much more functionality than previous versions of HAL. It's is also considerably more expensive ($369) than previous versions of HAL. But for the extra money, you truly do get far more functionality.

This screen allows you to view and edit the Reminders that you have created in HAL.

Date	Time	Type	Processed	Notfmethod	Notfdata2
02/07/01	16:59	REMINDER	X	SPEAKER	Don't forget doctor's
02/21/01	21:23	REMINDER	X	SPEAKER	Pick up Mandy.
03/29/01	08:15	REMINDER	X	SPEAKER	Take out garbage.

Reminders Add Remove Done

Figure 10-1 *Use the Reminders screen to view and edit any reminders that you have set up or to access the screens to set up new reminders.*

Now, let's set up a reminder.

1. Begin by right clicking on the phone icon and selecting Reminders from the menu that appears. This will open the Reminders screen, shown in Figure 10-1.
2. Click Add to add a new reminder. The Reminder screen, shown in Figure 10-2 appears.
3. To set up a new reminder, click on a day in the calendar to add a reminder to.
4. Next, in the Text Message field, type the words that you want HAL to say when the reminder runs.
5. Now, if you want a sound to also play when the reminder runs, place a check mark in the box next to Play Sound.
6. Then select Wav File to choose an existing WAV (sound) file for HAL to play or to select Rec Wav to record your own message or sound using the computer's microphone. If you select the Wav File option, you'll be prompted to open the file

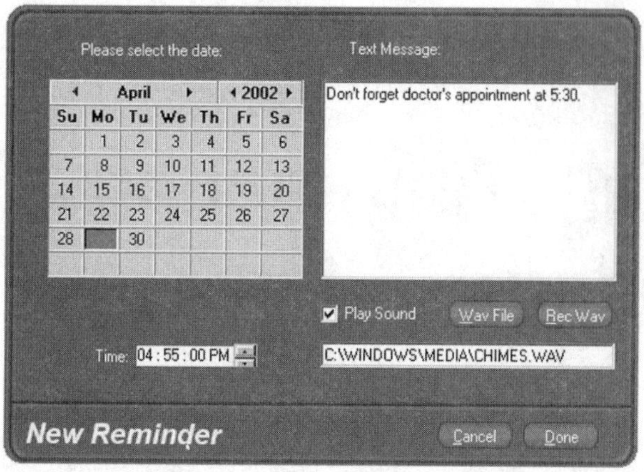

Figure 10-2 *Use the New Reminder screen to create new reminders.*

that you want to be delayed. Navigate to the file and then double click it to add it to the reminder.

7. You should see the path to that file displayed in the field below.
8. Finally, select a time for the reminder using the small up and down arrows next to the time field. Alternatively, you can simply highlight the existing time and replace it by typing a time into the field.

When you've finished making your selections, click the Done button at the bottom of the screen, and your reminder will be created.

> **Note**
>
> You can also set HAL2000 up to work with Microsoft Outlook. When you do that, you can create your reminders in Outlook and they will automatically be activated in the HAL2000 program. If you need recurrent reminders, it's best to set them up in Outlook.

Calendaring Applications

When you're creating reminders of any type, they're usually associated with some kind of schedule, which you usually track with a calendaring application. If you're using a web-based calendaring application like Google's Calendar, you may find that it's difficult to integrate that calendar with your home automation software.

There are some calendaring applications, however, that do integrate with home automation projects pretty well. In the case of this project, where you're programming HAL to remind you to do something based on a specific schedule, a calendar application, like Microsoft Outlook will make scheduling the regular event much easier.

HAL does integrate with some calendaring applications, just as some of the other home automation management software also integrates with calendar applications. Usually, you have the option to use a built-in calendar application or choose one of those that's compatible with the management software you've installed.

When you're integrating existing calendars, the most frequently used calendars for home automation are Microsoft Outlook, and iCal. These two applications offer both calendaring functionality and integration with some of the most commonly used home automation projects.

The calendaring application in Microsoft Outlook is actually just a portion of the program. In addition to calendaring, there's also an e-mail application and several other programs that make up the complete Outlook program. The e-mail application and calendar are two of the most widely used Microsoft programs.

Because it is so widely used, Microsoft Outlook integrates seamlessly with some home automation management programs, like HAL. Using Outlook, you can schedule reminders and specific actions in HAL.

iCal is a calendaring application that's similar to Outlook, though it's a Mac-based program, rather than Windows-based. For that reason, fewer management programs will integrate with iCal, and the ones that do will likely be Mac or Linux-based.

There are exceptions to that, however. HAL, for example, works on either Windows or Mac and so integrates with either Outlook or iCal. In general, however, these will be the two most commonly compatible calendaring applications, and using them will give you additional control and automation capabilities with your home automation system.

Moving On

So now you know. Yes, your house can talk to you. Or, more correctly HAL can talk to you. And it can remind you that you have appointments coming up, that the trash needs to go out, or even that you have a date of Friday night. All you have to do is tell it to remind you.

Project 11

No More Dead Lawns: Automating Watering Systems

Equipment Needed

- Orbit Automatic Yard Watering System
- Screwdriver
- Four Watering Hoses with Sprinklers Attached

Watering your yard and outdoor plants can be a real chore. It requires that first you remember to do it, and then that you take time to go outside and place the sprinkler and then move it. And installing a fully automatic watering system is expensive. Not only that, but a fully automatic watering system also requires that your yard be dug up for pipe placement, right? Well, not exactly.

Actually, installing automatic watering systems doesn't have to cost a fortune and it doesn't have to be a big dramatic project. You can install an automatic watering system for under $100 and it should only take a few minutes of your time. Don't believe it? Read on.

For this project, we're going to use an Orbit Automatic Yard Watering System. You can probably pick one up at any local home improvement store. The project is two parts: first you need to install the timer and connect the hoses and then you need to program the timer. Easy. Really.

Installing the Automatic Yard Watering System

1. Begin by installing the brass manifold that comes in the kit. Remove it from the package and attach it to the desired outdoor faucet by screwing it into place, tightly. Make sure that all of the water valves on the manifold are in the off position.
2. Next, connect the four valves that were provided in the package. These valves will control the on and off mechanisms of your hoses. The valves simply screw into place, just as a hose would.
3. Now, connect hoses to each of the four valves and place the hoses (attached to sprinklers) in the desired locations in your yard.
4. Using the hardware provided, mount the timer on the wall, near enough to the valves to allow connection to the timer. (If you don't want to mount the timer to the wall, you can also mount it on the bottom of one of the valves.)
5. Once you have everything installed, connect the wires attached to the valves to ports on the bottom of the timer. Be sure to match the wires to the correct ports, from left to right. (For example, the farthest valve on the right should be connected to the farthest port on the right.)
6. Finally, open the water valves on the brass manifold and slowly turn the water on to ensure that there are no leaks in the connections.

The automatic watering system is now connected. All you need to do now is program the system to water at the times and for the durations desired.

Programming the Automatic Watering System

Once you get the watering system installed you need to program it to turn on and off automatically. This programming is a four step process that includes setting up the timer with the correct time and day, setting watering days, setting the start time and duration of watering, and enabling automatic watering.

Setting Up the Timer

Before you can begin to program the timer, you need to install three AA batteries. The battery compartment is located beneath the timer cover.

1. Once the batteries are installed, press the Reset button on the lower right side of the unit. The button is slightly indented, so you may need a tool (such as a pen) to reset it.
2. The unit will go into the Set mode, as indicated by the word SET blinking on the screen.
3. Press the Confirm button to confirm that you want to set the unit up.
4. The word CLOCK will appear on the display. This indicates that you are setting the day and time. Push the plus or minus buttons to select the day of the week.
5. When you've set the day of the week, press the Confirm button to accept. Then the time will begin to flash.
6. Again, use the plus and minus buttons to set the clock to the current time. Press confirm when the time is set to begin setting the watering days.

Setting Watering Days

1. If you're setting the watering days immediately after setting the correct day and time on the unit, then press the plus or minus buttons to select the next day or second or third day watering schedules.
2. When you reach the desired schedule, push the key with two arrows (in the right corner) to toggle on watering, which is indicated by a small rain drop.
3. When your watering schedule is set, push the Next button to move to the duration and start time settings.

Note

If you're setting a schedule, but you did not just set the correct time and day, you can enter into the watering days schedule by pressing the Confirm button three times. Then select your watering days as detailed above.

Setting Duration and Start Time

The duration and start time should be set for each of the four water stations (outlets to which hoses are connected).

1. If you're setting the duration and start time immediately after setting the watering days schedule, then push the plus or minus buttons to set the duration of watering (1 to 99 minutes) desired for the first station.
2. After you have selected the desired duration, press the confirm button to advance to the duration for each of the next stations in turn until all four stations have been set.
3. When you press confirm after the fourth station has been set, you'll be taken to the screen for setting the start times for watering.
4. There are three watering cycles. You can set a start time for any one or all three cycles by pressing the plus or minus key to adjust the start time.
5. When the start time is set, pressing the confirm button again will move you to the next cycle.
6. After you have set the start time for any (or all) of the cycles, press the confirm key one final time. This will return you to the Mode selection.

Note

If you're setting the duration and start time, but you did not just set the schedule, you can enter into the duration and start time programming

(Continued)

(*Continued*)

screens by pressing the Confirm button three times and then pressing the Next button once. Then select your duration and start time scheduling as detailed above.

Enabling Automatic Watering

Automatic watering may already be enabled on your unit when you complete the program. If it is, the Auto indicator at the top of the display will appear. If not, you can enable automatic watering by pushing two buttons.

To enable automatic watering, press the Mode button. The word AUTO will blink on the screen. Press the confirm button to begin the automatic watering schedule.

Alternatives to Windows-based Automation

If you're going to include computer-based automation in your home automation system (like much of the X-10 automation capabilities) but you're not a Windows user, you might feel a little left out in the cold.

For sure, the majority of home automation software applications are Windows-based. That's because the majority of computer users in the world are Window-based users. However, if you don't fall into that majority, you're not completely left out. There are some excellent applications for both Mac and Linux operating systems.

- **XTension**: (http://www.shed.com/) The most well known (and probably the most useful) software program for home automation on the Mac platform is the XTension application. This application allows you to control your X-10 devices and because it has AppleScript support, you can customize the application to meet all of your home automation control needs.
- **Indigo**: (http://www.perceptiveautomation.com/indigo/index.html) Another Mac-based application that's very usable is Perceptive Automation's Indigo program. Like XTension, Indigo is AppleScript-enabled. However, if you're not the kind of Mac user that loves tinkering with the code, you can use this software right out of the box to control a variety of X-10 devices.
- **MouseHouse**: (http://www.mousehouse.net/) MouseHouse software is another Mac-based home automation control software. While it's not quite as powerful as XTension or Indigo, MouseHouse is a good software for home automation beginners, and it is customizable if the features that you want aren't included.
- **Xtend**: (http://home.comcast.net/~ncherry/common/xtend.tgz) Xtend is a Linux-based home automation software, however, it only works to execute commands, not to receive them. To create a complete solution, Xtend should probably be used with another Linux-based software.
- **FlipIt**: (http://www.lickey.com/flipit/) FlipIt is a simple command line home automation control program that can be used with Unix-based home automation devices and computers. This is a home-grown program, however, so support for it is limited, and it requires a good bit of programming knowledge to be truly useful.
- **BlueLava**: (http://directory.fsf.org/bluelava.html) BlueLava is a CGI-based home automation software. It allows you to control X-10 modules and devices from within a web browser.

See, there are plenty of options, depending on how much time and effort you want to put into the management software that you select. If you're a programmer (or even a hobbyist that happens to be very good at programming), you can use some of these programs to extend control far beyond the norm.

If, however, programming isn't your thing, then simple, easy to use software applications for Mac and Linux-based home automation also exist. You get to decide what you like. And if nothing here appeals to you, search the Internet. There are many other programs available.

Moving On

That's all there is to it. You don't have to think about watering your outdoor plants ever again. Simply set up an automatic watering system and the timer will do all of the work for you. Now if only kids and pets were that easy!

I can't think of any way to automatically feed your kids that doesn't require some regular attention from you, but I can show you how to keep an eye on them (and your pets) if you're not home when they are. It's in the next chapter.

Project 12

Watching from Afar: Monitoring Kids and Pets from Work

Equipment Needed

- Veo Wireless Observer One Network Camera
- Windows-based PC
- Highspeed Internet Access (Cable/DSL)
- Networking Cables

If you're a parent, then you know that keeping up with your kids can be difficult, especially if they're home when you're not. Wouldn't it be nice if you could keep a (discreet) eye on them, even when you're not there?

Maybe instead of kids, it's your pet that you want to keep an eye on. Either way, you can. A webcam that you can access from the Internet is the perfect solution. And it won't cost you a fortune to install.

In this project, we're going to use the Veo Wireless Observer One Network Camera and configure it so you can access live video feed in real-time using any web browser. This is a three part project: installing the hardware, installing and configuring the software, and accessing the webcam from the Internet.

Installing the Hardware

So you've decided you want to keep an eye on things from afar? No problem. Installing the Wireless Webcam from Veo is easy enough.

1. First, plug the networking cable that came with the webcam into the camera. Then, plug the opposite end into your cable or DSL router.
2. Next, connect the power adapter to the camera and then plug it into the wall.
3. Finally, press the power button on the base of the camera.

The unit will take about 30 seconds to power on and locate its IP (Internet Protocol) address. The last three digits of that address will be displayed on the base. Make note of those three digits, as you'll need them to access the camera from your web browser.

Now your camera is ready for action. The next step: installing the software.

Installing the Webcam Software

Once you get your camera connected to your cable or DSL router, then it's time to install the software that will enable the camera and allow you to access the video feed from the Internet.

Begin by inserting the software disk into your CD drive. The CD should auto-run. If it does not, you can start the installation by going to Start > Run and then click browse to navigate to your CD drive. Select the drive and then locate the AutoRun.exe file. Double click the file and the software will begin the installation process.

Click the Install Observer XT Setup Utility to install the software.

When the installation is complete, click the Check System Requirements Utility link to check that you have the minimum system requirements to run the system. If you do not, you can download the latest updates from the Internet.

When the Check System Requirements Utility is finished, click Exit.

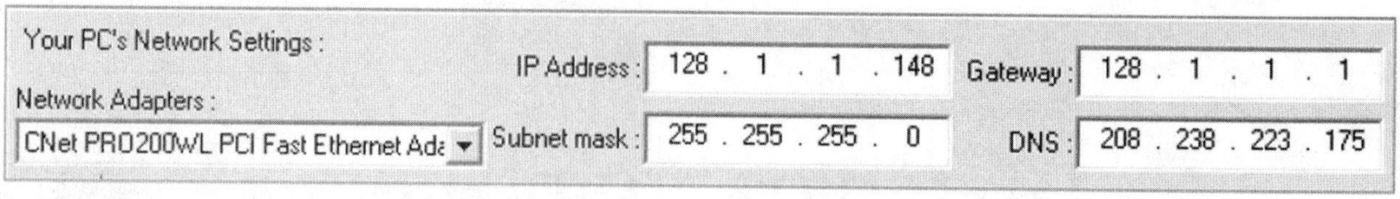

Your PC's Network Settings :
Network Adapters : CNet PRO200WL PCI Fast Ethernet Ad
IP Address : 128 . 1 . 1 . 148 Gateway : 128 . 1 . 1 . 1
Subnet mask : 255 . 255 . 255 . 0 DNS : 208 . 238 . 223 . 175

Figure 12-1 *The Observer XT setup screen.*

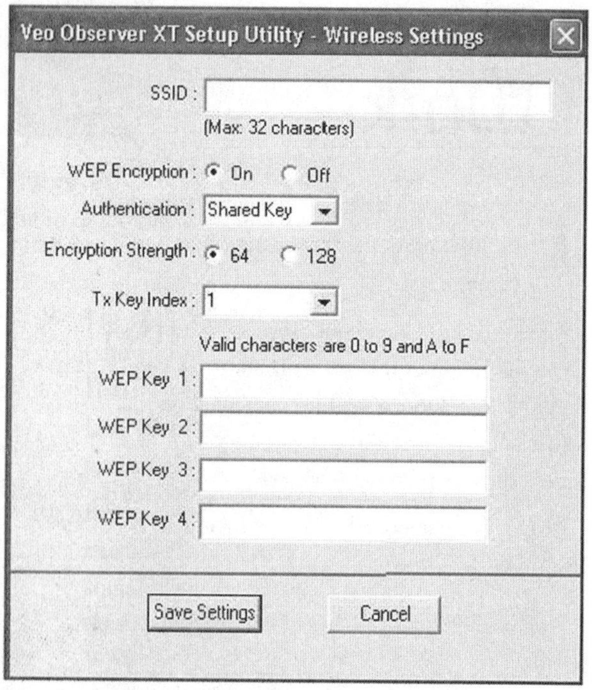

Veo Observer XT Setup Utility - Wireless Settings
SSID :
(Max: 32 characters)
WEP Encryption : On Off
Authentication : Shared Key
Encryption Strength : 64 128
Tx Key Index : 1
Valid characters are 0 to 9 and A to F
WEP Key 1 :
WEP Key 2 :
WEP Key 3 :
WEP Key 4 :
Save Settings Cancel

Figure 12-2 *Use the Wireless Setup screen to ensure that all of your settings are correct.*

Configuring the Software

Once you have the software installed, you need to configure it to work with the camera and for web access.

1. To access the setup utility, double click the Observer XT Setup Utility icon on your desktop. You'll be taken to the setup screen.
2. The first section on the setup screen is Your PC's Network Settings, as shown in Figure 12-1. The information displayed on this screen is your current network information.
3. The second section on the screen is labeled Cameras on Local Area Network. In this section, select your camera and then click the Wireless Settings option.
4. You'll be prompted to enter your username and password. After you enter them, click OK.

> **Note**
>
> Unless you have changed the username and password, the default username is "admin" and the default password is "password." You can change those settings on the User Management Screen.

5. The Wireless Setup screen shown in Figure 12-2 will appear. On the screen, your current wireless settings should be displayed. These settings should be the same as the settings for your wireless access point or router in order for the camera to operate properly.
6. The settings include:
   - SSID (Service Set Identifier) is a unique name designated for a specific wireless local area network. Enter the SSID of the access point or router that you'll use to access the camera. The SSID used by the camera should match your wireless access point or router's SSID. It is case sensitive and cannot exceed 32 characters.
   - WEP (Wired Equivalent Privacy) is a wireless security protocol that encrypts the data sent over the wireless network. WEP encryption is set to Off by default. If you have enabled WEP on your wireless access point or router, select On.
7. If you enable WEP, you will also need to select whether your wireless access point uses 64 or 128 bit encryption.
8. Select the Tx Key Index (transmit key index) number that your wireless access point or router uses. This number indicates which of the four WEP keys is currently enabled.
9. Finally, enter your WEP key into the corresponding field. When you've entered

the required information, click Save Settings.

10. Next, turn the camera off and disconnect the Ethernet from the back of it. After it is disconnected, turn the camera back on. It should take about 30 seconds for the camera to start back up.
11. To ensure that you've configured the camera correctly, go back to the Setup Utility and click Refresh List. If your camera appears on the list, you've configured it properly. If the camera does not appear on the list, you'll need to go through the setup steps again to ensure that you configured it properly.
12. Back on the Setup Utility page your Camera Network Settings should be set. If they are not, then click the radio button next to DHCP (Dynamic Host Configuration Protocol) to allow the IP address and other network settings of your camera to be automatically configured. Make note of the camera's IP address, as you'll need this address to access the camera from the Internet.
13. When you've adjusted these settings, click Save and you will be prompted to enter an administrator name and password to confirm the settings.
14. Once this is complete, your camera will automatically restart. It could take as much as 30 seconds for the camera to come back online, so be patient.
15. When the camera has reset, you can click the Login button to launch your web browser and go to the camera's login page in the event that you need to change some aspect of the camera's setup.

Now that you have the camera set up, you can access it from anywhere via the Internet using a web browser.

Accessing the Webcam from the Internet

> **Note**
>
> To access your camera from the Internet your web browser needs to support XML and Java. If your browser does not support these utilities, you will only have access to JPEG Refresh Mode, which is still pictures.

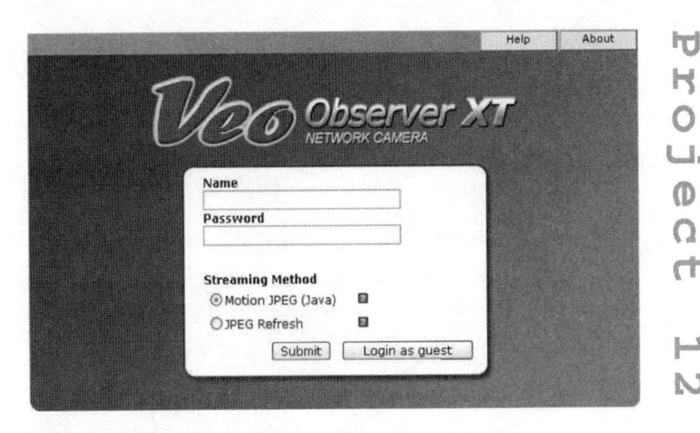

Figure 12-3 *When you type your camera's address into your browser's address bar you'll be taken to the login screen.*

To access your camera from the web, open your web browser and type the camera's IP address into the address bar. It should look something like this: http://168.168.1.123. The first three sets of numbers should be the same as all of the other equipment on your network. Only the last three numbers will be different. These three numbers are the same numbers that are displayed on the base of the camera.

You should be taken to the login page for your camera, as shown in Figure 12-3.

Enter your username and password and select how you want to view the video (as a stream or refreshing still pictures) and then click submit.

After you click submit, you may be prompted to install a Java applet before the video can start. If you are prompted, click Yes. The applet will install and your video should begin streaming within a few seconds.

That's it! You now have the ability to check in on your kids or your pets from anywhere, and the camera has a range of about 295 feet, so it should work anywhere within your home.

Choosing the Best Webcam

When you begin to think about the types of webcams that you might want to have connected to your home automation system, it would seem that all webcams are created equal. In fact, they are not. There's much to learn about webcams, and this intro won't cover it all, but you'll get the basics, and then if you want to learn more there's plenty of information on the web or in books.

To start with, the way and place that you intend to mount the webcam that you'll be using will make a difference in the type of webcam that you should purchase. For example, if you plan to connect your webcam directly to your flat panel monitor, then you'll need one that's designed to clip to the monitor. If you plan to mount your webcam outside, or on a flat spot in your house, then you'll need one that can be screwed into the surface on which you plan to mount it.

Cost is always a concern as well. There are many different types and prices of webcams available, all with different specifications and ranges. Examine them carefully to be sure that you're getting a webcam that will perform the way that you expect for it to perform within the range that you plan to use it. For example, a webcam used for video conferencing will have different capabilities than a live webcam for watching children, pets, or outdoor life.

You should also keep the requirements for the camera you decide on in mind. Some have very specific input requirements (like USB 2.0), graphics card requirements, and RAM (Random Access Memory). Be sure that the camera you select does not require more resources than you have available, and if you're running an older model computer, consider upgrading before you begin your home automation projects.

Here are some additional questions that should help you as you search for the right webcam for your home automation projects:

- Who will be watching the feed from your webcam? If it's just you, then you may not need to invest as much in the camera. However, if you're setting up a camera that you want to allow the general public to view through the Internet, then you should invest a little more for a higher quality camera.
- How much detail do you want to be able to see? If your camera is for security purposes you may want more definition than if you're just monitoring your children or your pets.
- Will you use the camera outdoors or indoors? Outdoor systems have to be weather proof and you have to consider the sun position. Lighting is also a concern that you may want to take into consideration, because some webcams perform very poorly in low-light conditions. If you're putting a camera in a birdbox, for example, you'll want to be sure that it captures good low-light pictures, because there will be little light inside the birdbox.
- What wiring is available in the location where you want to mount the camera? Consider both network cabling and power requirements. If you plan to put the camera in an area where wiring is unavailable, then a wireless, battery operated camera will be your best choice.
- How large is the area that you want to record? The type of lens (and the moveability of the lens), is determined by the area that you're interested in seeing. For example, the birdbox cam will be recording a much smaller area than a camera that's recording your children in the living room or their bedrooms.

There are other considerations when you're selecting webcams, too. For example, do you want to archive the videos that you're recording, or do you want to skip recording completely and just provide a live webcam feed? These are all elements of connecting and using webcams that should be taken into consideration, but to get you started, the information above is a great jumping-off point.

If you're interested in learning more about webcams and the various aspects that you should take into consideration, you can learn more by using Google to search for information, or you can find great books on webcams at your local bookstore. (I highly recommend *The Little Webcam Book*, by Elisabeth Parker if you can find a copy of it.)

Moving On

Who knew it could be so easy to keep an eye on your kids and pets even when you're not around? The Observer camera is the perfect option because it's easy to install and access, and it won't break the bank. And with the wireless option, you're not tied to having the camera only in the same room as the computer.

Project 13

Hot and Ready: Preheat the Oven from Your Cell Phone

Equipment Needed

- Connect Io Intelligent Oven

Have you ever had one of those nights when you were on your way home, stuck in traffic while dinner just gets further and further behind? Wouldn't it be nice if you could call your oven and tell it to preheat while you were cooling it off on the freeway? You can. But it will set you back a pretty penny.

The Connect Io Intelligent Oven is not only compatible with your telephone lines, you can also access it via the Internet or set it up on a timer (or recurring timer) to turn itself on and heat up at a given time of day, on any given day. It's the home-cook's dream come true.

It's also about $4,000 and that doesn't include the cost of installation. Still, wouldn't it be fun to have one? We're going to learn about programming one in this chapter, just to have something to work toward in the future.

Installing the Oven

Most electrical installations should be done by a trained professional. Installing an oven is no different. Except that this is no ordinary oven. Not only are there voice control capabilities and web-enabling features, the Connect Io Intelligent Oven also has refrigeration capabilities. That makes this oven very different from most ovens, so it's best if the installation of this appliance is left up to the professionals. Instead of installing the oven yourself, have someone else do it, and we'll show you how to gain access to it by the phone.

Programming the Oven

The Connect Io Intelligent Oven comes with a variety of features, including the ability to be accessed via your home network, and by phone. We'll concentrate on the phone feature in this project.

Once you've had your oven installed, you can set it up to be active remotely. Unless you set up the remote activation on the unit, however, you won't be able to call in to it.

1. The first step in setting up the oven then is to enable remote access. To do this select My Oven from the drop down menu on the main screen.
2. Select Remote Access.
3. Then select Phone # and enter the 10 digit phone number you are most likely to call from.
4. Next, press Pin # and enter a four digit code that you will be able to remember.
5. Push the Remote Access button to toggle it on.
6. Finally, select OK to save your settings and return to the main screen.

Now your oven is prepared for you to call it and give it instructions. When you call the oven, the number you are calling from will automatically be detected by the oven's caller ID function. If you're calling from the number you entered above, you'll be prompted to enter the pin number you chose. If you're calling from a different phone, you'll be prompted to enter the 10 digit phone number you registered in the Remote Access configuration.

Once you've enabled Remote Access on the oven, you can control it by phone by calling 1-919-882-2330. When the call is answered, you'll be prompted to enter your phone number (if calling from a non-registered phone) and pin number. After you've entered the correct number, you'll be given options for controlling the oven or checking the temperature.

When you call your oven, you can communicate with it by speaking into the phone or by pressing the appropriate keys on your phone's touch pad.

You May Already Have X-10-Enabled Devices

Remember how we said that X-10 is a protocol *and* a technology? Well, guess what? It's a protocol and a technology that you might find lurking in some of the other devices that you already have floating around your home.

X-10 is not a new technology. In fact, X-10 has been around since the 70s and since that time, it's been quietly added to a lot of the devices that you wouldn't ordinarily think about. And it's possible that you have some of these X-10-enabled devices in your home.

One of the ways that you can tell if a device is X-10 enabled is to look at the manufacturer or name brand on the product. Because the company X-10 Incorporated makes devices for so many other companies, the technology might be included and you just never knew it. Here's a list of some of the name brands that probably contain X-10-technology:

Advanced Control Technologies

GE Homeminder

HomeLink

HomePro

IBM Home Director

Leviton Manufacturing Company

Magnavox

NuTone

PCS

RadioShack (Plug 'n Power Brand)

RCA

Safety First

Sears

SmartLinc

Stanley

Universal Electronics

Wesclox

X10 Powerhouse

So, how do you tell if you have a device that carries one of these name brands that's X-10 enabled? Well, in some cases you'll find them labeled as being X-10 enabled. In other cases, however, there is no such indication.

If you find no indication on the device that it's X-10 enabled, then you may find that you can tell by looking at the device. Some contain the same basic design that all X-10-enabled devices do, with the ability to set a house code and device code. Or, you'll find that the device is remote (or infrared signal) controlled. These are the most common ways to tell, but you won't always be able to find the answer just by looking at the device.

If you can't tell by inspecting it, try reading the owner's manual. There will most likely be some notation about it within the pages of the manual. If you don't have the owner's manual, you can always call the company.

X-10 is a far more commonly used protocol than you might think. Because it's been around for so long, it's been quietly slipped into many of the devices that you use around your home. So, without even knowing it you might already be well on your way to creating a connected, X-10-enabled smarthome.

Moving On

Currently there just aren't many good options for controlling your oven remotely. That's mostly because of the complexity of the appliance. However, over time more and more remotely-controllable appliances, including ovens, should come on to the market. Until then, the Intelligent Oven is a good choice for accessing your oven remotely.

Even the cost of this oven isn't all that overwhelming when you consider the hours of time and frustration you'll save by having the ability to control the oven remotely. And that doesn't even take into consideration that the oven also has refrigeration capabilities.

Project 14

Who Let the Dogs Out: Creating Your Own Robo-Dog

Equipment Needed

- Hal2000
- Motion Sensor
- WAV File of Dog Barking

If you live alone there are times when you just don't feel safe, especially if you're a single woman. If that's the case, you've probably considered getting a dog. Man's best friend is the perfect solution to being alone. It can bark and scare off intruders or warn you if there's someone at the door. Even families sometimes wish they had a dog to keep them safe.

The problem with dogs is that they require a lot of maintenance and care. There's the constant walking, the regular feeding, bathing, trips to the vet, and even playtime to take into consideration. It takes time to care for a dog the way they need to be cared for. And that's to say nothing of the restrictions that are placed on pets if you live in a rental home or apartment. There has to be a better way to have the security of a dog without all of the hassle.

There is, and it's in HAL2000. HAL2000 has some great voice and sound capabilities that make it perfect for creating illusions to keep you safe. And that's what we'll do in this project. Using a motion sensor and HAL2000, you can set up controls so that if the motion sensor is triggered, a dog will begin barking, making it appear that you have a protector keeping you safe.

In previous chapters you've learned how to install wireless motion sensors, so we're not going to cover that again in this chapter. Instead, we'll focus on setting up an If/Then situation with HAL that will result in the barking dog WAV file playing if your motion sensor is tripped. So, be sure you place your motion sensor in an appropriate place – where it can detect someone approaching your front or back door might be appropriate – and then add it as a device to the HAL program so that HAL is monitoring the sensor. Then you can program the barking dog response.

Programming Your "Dog"

After you've installed your motion sensor you need to add it as a device in the HAL program, then you can create an If/Then situation that tells HAL, "If the motion sensor is triggered, then play this WAV file."

1. Begin by opening the Automation Setup screen, as shown in Figure 14-1, and click the Tasks button.
2. On the Tasks screen, shown in Figure 14-2, click the If/Then Situations (Rules) option.
3. Then click Add at the bottom of the screen to open the Rule Add Wizard screen.

Figure 14-1 *The Automation Setup screen is the starting point for creating your dog.*

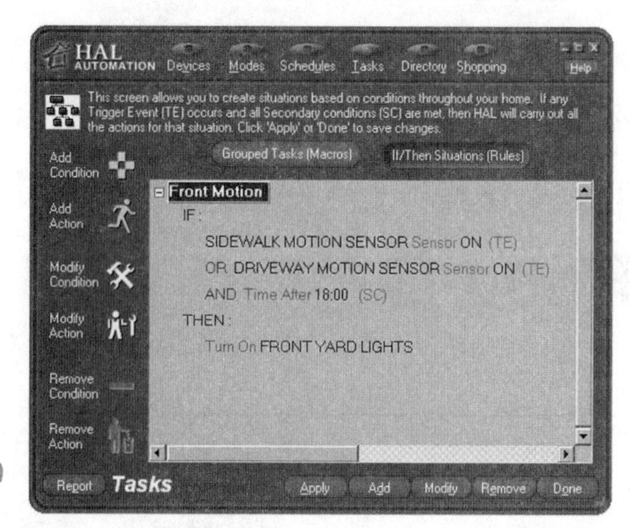

Figure 14-2 *Use the Task screen to create a new If/Then situation.*

Figure 14-3 *Give your new rule a name.*

4. As shown in Figure 14-3, you'll be prompted to give the rule a name. Type the name you want it to have and click OK.
5. The Conditions Wizard should appear. Select Trigger event, because the motion sensor will trigger the event you want to take place, and then click Next.
6. On the Conditions screen that appears (shown in Figure 14-4), select sensor from the Condition dropdown menu, since you're setting up a sensor. Then select On for The condition is met when the sensor…and select the sensor from the list below. When you've made your selection, click OK to add the condition and close the Conditions Wizard screen.
7. You should be returned to the Modes screen. From there click Add Action and the Action Wizard, shown in Figure 14-5, will appear.
8. In the center of the screen, left click on a device to select it.
9. Then in the Action drop-down menu near the top of the page, select Play WAV File. The WAV File Action Wizard, shown in Figure 15-6, will appear.

Figure 14-4 *Use the Conditions screen to set the conditions for the event you're programming.*

Figure 14-5 *The Action Wizard controls what action will take place when the motion sensor is triggered.*

10. Using the drop down menu on the left side of the center of the screen, select the location where your WAV file is stored.
11. Then on the right side of the center of the screen highlight the specific WAV file you want to use by clicking the name of the file, and then click OK.
12. You will be returned to the main Action Wizard screen.

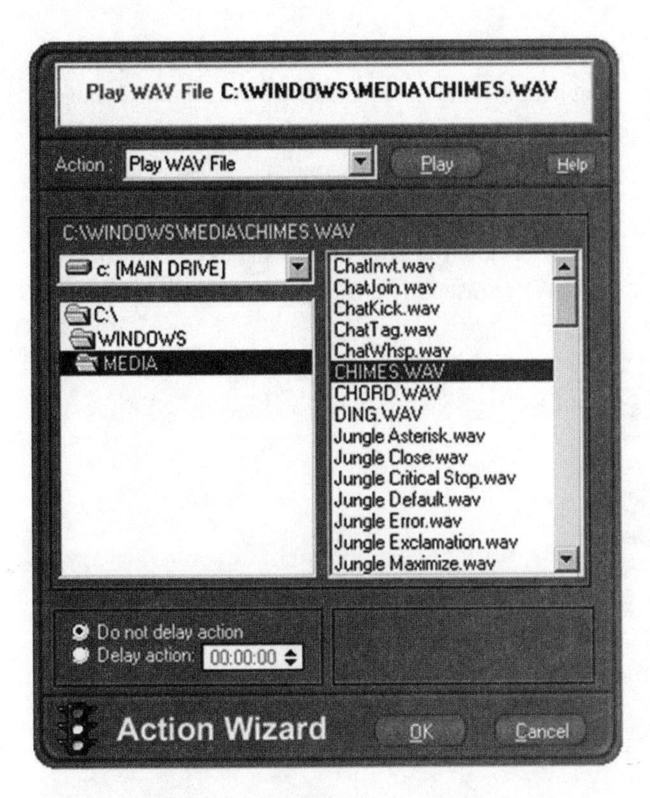

Figure 14-6 *Use the WAV File Action Wizard to select the barking dog WAV file.*

13. Now, click the Action radio button to set the action to On.
14. Then make sure the radio buttons next to Do not delay action and Do not confirm this action are clicked.
15. Click OK to save the action and close the Action Wizard.

Your Robo-Dog is ready to come to your defense. To ensure that the If/Then rule is working properly, set the motion detector and then trip it. The If/Then rule you just set up should trigger your WAV file to play through whatever speakers you have attached to your computer. You can adjust the volume as necessary.

Should You Network Your Home and Computer?

Many of the projects that are covered in this book just happen to be computer-based projects. And you might be wondering if you should network you home with your computer. The answer is, well…probably.

Don't groan too loud. You can start with a simple network that connects together your home computers, laptops, and entertainment equipment. That type of network would consist of a router and some cabling. You might even need a print router, but if you connect the network properly and allow file and print sharing on your network you probably won't need a print router.

At some point in the future, and depending on how much home automation that you want to implement (and the kinds of automation that you want to implement) you might decide to add your own server to the mix. This is a pretty big undertaking though, so you should not jump right into it without first learning more about it.

If you're interested in connecting your home and your network, you might want a few good books on the subject. Here are a few to get you started, arranged from the most basic to the most complex:

- *Home Networking for Dummies*, by Kathy Ivens, For Dummies Press, ISBN: 0764588494.
- *Home Networking: A Visual Do-It-Yourself Guide,* by Brian Underdahl, Cisco Press, ISBN: 1587201275.
- *The Rational Guide to Small Office & Home Networking*, by Jim Boyce, Mann Publishing, ISBN: 097268882X.
- *PC Magazine Guide to Home Networking*, by Les Freed, Wiley & Sons Publishing, ISBN: 076454473X.
- *Home Networking (The Missing Manual Series)*, by Scott Lowe, Pogue Press, ISBN: 059600558X.

Home networking on the simplest level is point-and-click easy. Today's home network routers usually come with a software tutorial that walks you through every step of the installation. However, when you start talking about networking a smarthouse, you throw more elements into the network than just computers and the Internet.

An automated home network might include networking computers, printers, laptops, televisions, sound systems, and even some appliances. It can be a very complex and confusing network to create.

If you're not certain that you possess (or can gain) the knowledge needed to create this whole-house type network, then you might want to bring a professional network or home automation installer into the mix. There are numerous installers listed in Appendix B, if that's the route you choose to go.

The benefit of having a home network that includes all of your automated devices is the ability to control all of those devices via a remote computer access, or a web-based application. It's not the easiest installation that you've considered, but it will be one of the most valuable.

Moving On

All of the HAL programs give you so many options for controlling your home and devices within your home that you're limited in what you can do only by your imagination. And now that you've programmed your motion sensor to play a barking dog WAV file, you don't have to worry about someone sneaking up on you while you're deciding on the next great project.

Of course, if you're not sure what to do next, flip the page. We're installing a Bird Feeder Cam in the next chapter.

Project 15

For the Birds: Installing a Bird Feeder Cam

Equipment Needed

- Security Man Outdoor Watch Wireless Camera Kit
- Post (or Tree) for Mounting
- Quikrete
- PVC Pipe (1/2 to 1 inch in diameter)
- Installed Bird Feeder
- Screwdriver
- Television

Birds are funny little creatures. They play and socialize and even communicate in the most interesting ways. And many people enjoy watching them so much that they install bird feeders or houses to draw the birds into their yard.

These folks are rewarded by birds of every color coming around to eat and play. The problem is, to watch the birds you have to sit in the house and watch through the window. Birds are skittish, so they usually take off as soon as any movement or noise breaks the normal atmosphere.

If you really want to see a bird up close and personal, you need to have a way to watch them without disturbing them. Watching through the window works, but it doesn't let you get close.

What will let you get close though is a wireless outdoor camera. And in this project, we're going to install such a camera and mount it near enough to your bird feeder (or feeding area) that you can get a good look at those birds that you've worked so hard to draw into your yard.

We're assuming that you already have bird feeders installed in your yard, so this project won't address installing them. If you don't have bird feeders, check with your local home improvement store. The people who work there can probably give you some excellent tips for placement and type of feeder for your area.

Installing the Camera

In this project, we're using the Security Man Outdoor Watch Wireless Camera kit, shown in Figure 15-1. This kit comes with everything you need to hook up your camera, except for a post to mount the camera on if your bird feeder is in the center of your yard or away from any solid object to mount it on.

And the first thing you need to do is decide where you want to mount the camera. Keep in mind that the optimal viewing range for this camera is around 45 feet, and

Figure 15-1 *The Security Man Outdoor Watch Wireless Camera kit has all of the equipment you need to set up your first bird cam.*

obviously the farther it is away from the bird feeder the harder it will be to see the birds. I would suggest mounting the camera about 10 to 15 feet from your bird feeder.

If there is no tree or wall on which you can mount the camera, then you'll need to install the post for mounting purposes. The viewing angle of the camera is 62 degrees, so you'll want to mount it slightly higher than your bird feeder.

1. To install the post, dig a post hole at the desired mounting location and mix Quikrete according to the manufacturer's directions.
2. Then, place your post into the hole, and fill the open area around the hole with Quikrete.
3. Use a level to make sure your post is set straight.
4. It could take several days for the Quikrete to dry completely, depending on the climate that you live in, so the installation of the camera needs to wait until the post is securely installed and the Quikrete is set completely.
5. Once you have a place to mount your camera, all you need to do to mount it is screw the base of the camera on to the post (or whatever you're using for a mount) at the desired level. Mounting screws should have been provided with the camera.

When installing the camera, be sure that you have a clear view of your birdfeeder and the area immediately surrounding it. There's nothing more frustrating than trying to watch birds when a tree, branch, or leaf keeps getting in the way.

Connecting the Power

The Security Man camera is a wireless camera, so you don't need to run any wiring to connect it to the base station. However, it is not battery powered, which means that you will have to connect the camera to an electrical outlet.

Because there's electricity involved, it's best if you run the power cord for the camera through a piece of PVC pipe. You can run the PVC pipe right up the side of the post to the camera and then on the ground bury the PVC pipe just below the surface of your lawn if you prefer. These aren't requirements, but they do protect the cord and ensure that you get the most life out of your wireless camera and receiver.

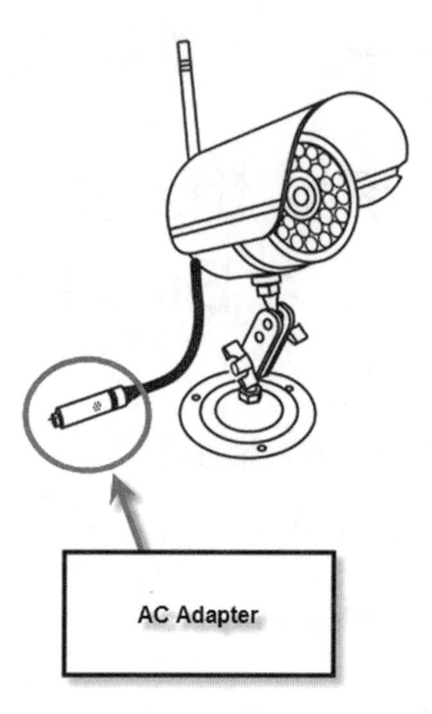

Figure 15-2 *Connect the power cord to the camera at the jack provided.*

1. To connect power to the camera, plug the AC adapter into the jack provided on the camera, as shown in Figure 15-2, and then plug it into a wall socket or extension cord, if necessary. Remember to keep the extension cord connection protected from the elements to reduce the risk of a fire or electrical problems.

Connecting the Receiver

Now that you've mounted and powered up your camera, you need to connect the receiver. The receiver will connect to any television or monitor and to some VCRs, DVD players, or DVR units via the RCA jacks provided. A connection diagram is shown in Figure 15-3 to help you visualize how the system will be connected.

Before you connect the receiver to your television (or other viewing method), you should connect the antennas provided to both the camera and the receiver. The antennas simply screw onto the connection provided.

Once you have your antennas connected, plug the RCA jacks into the television (note that they are color coded. Connect red-to-red and yellow-to-yellow). Then plug the opposite end into the correct spot on the receiver unit, as shown in Figure 15-4. Once you have

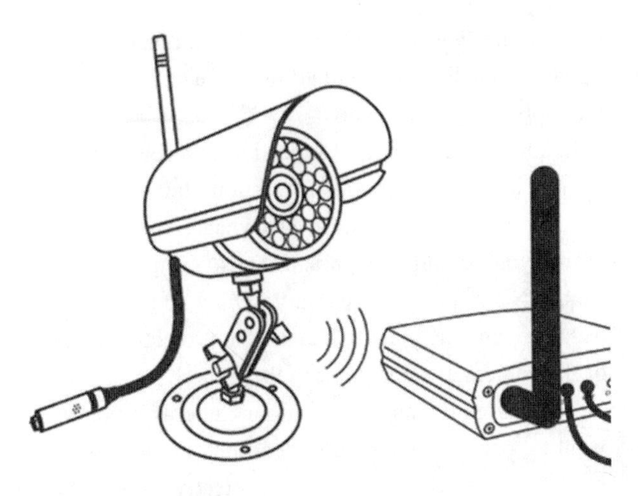

Figure 15-3 *Connect the receiver to the television to watch your birds with the wireless camera.*

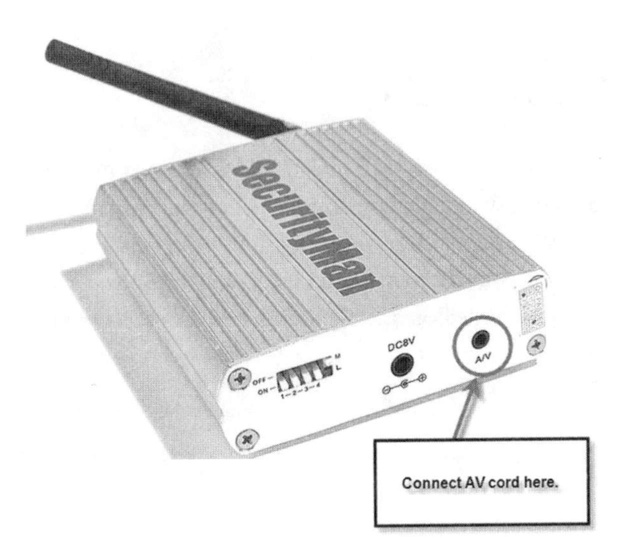

Figure 15-4 *Plug the provided RCA jacks to the computer and then to the receiver, matching the connecting wires.*

the receiver connected to the television, configuring it to receive your video is almost effortless.

Configuring the Receiver

When you unpack the wireless camera and receiver from the manufacturer, both pieces should be set to the same frequency, which means they should work well together, straight out of the box.

However, if you want to check that the devices are set alike, you can check the Dip Switches, which are

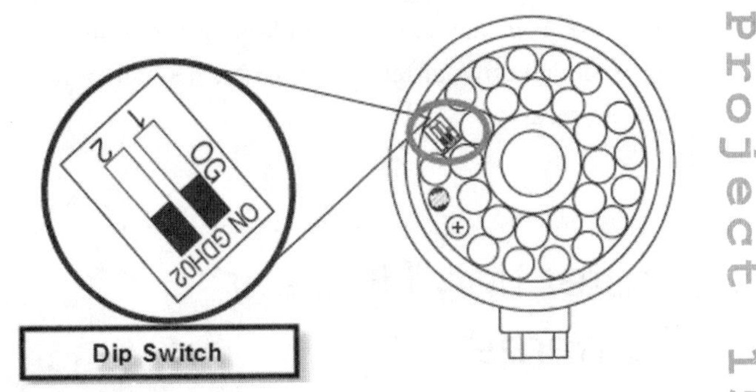

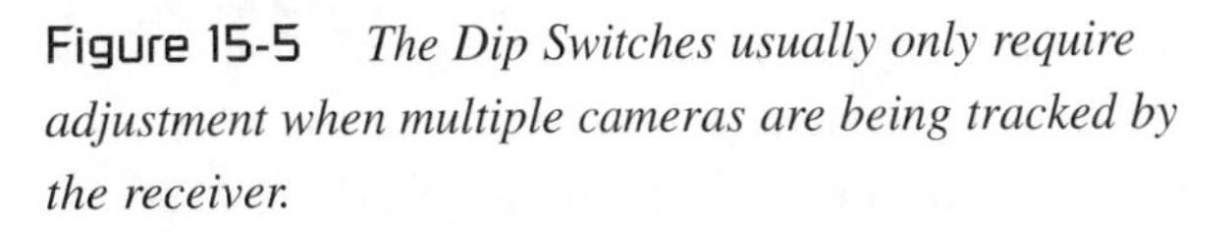

Figure 15-5 *The Dip Switches usually only require adjustment when multiple cameras are being tracked by the receiver.*

located on the font of the camera and on the front of the receiver. To get to the Dip Switch on the front of the camera, you have to screw the protective cover off the lens. Then, as shown in Figure 15-5, the switch is located right on the face of the camera.

1. Check to ensure that both switches are set to the same channel. If they're not, then the receiver won't be able to receive and relay video.
2. As long as these switches are set to the same setting, don't change them. The Dip Switch is intended to allow more than one camera to be used with each receiver.
3. Once you're certain that the switches are set the same, replace the cover on the front of the camera.
4. There is no further configuration needed for the receiver to begin receiving and transmitting video feed from the camera to the television. To find out if you've set the camera and receiver up properly select the television channel that the units are programmed to work on. This is indicated by the DIP switch that is turned on.
5. If the camera is working properly, then you should see the real-time video feed displayed on the television.

If, for some reason you're not receiving video, double check all of your connections and the Dip Switch settings to ensure that everyone is connected properly and that the units are keyed to the same channel.

One other benefit of this system is that you can add more cameras to it, if you decide that you would like to get different views of the activity that's taking place in your yard. You can connect up to four cameras, and view them all on your television, with the single wireless receiver.

Protecting Outdoor Cameras

One of the biggest issues that many people find when they're using cameras outside in home automation projects is that most of the cameras you purchase are not made for outdoor use. The ones that are made for outdoor use can be extremely expensive, and therefore they can reduce the number of cameras that you're willing to use outdoors.

You can, however, create weatherproof outdoor enclosures for your camera. Some people build these enclosures out of weather-treated wood, sealed with waterproof caulk or some other type of weatherproof sealant. Others use PVC pipe and weatherproof caulk.

Basically, what's important is that the camera and the wiring connected to the camera are protected from the weather. To do this, you'll want a design that allows the camera to clearly shoot pictures while remaining protected. In most cases, you can purchase pre-cut glass or Plexiglas that will do the trick. Then a bathtub sealant or caulk, applied to all of the seams will keep the moisture out of the box.

In some cases, you'll also need to ensure that the camera is steady. A good construction glue is an option if you can't mount the camera to whatever protective casing you've built. One brand that's very good is Liquid Nails. Just be sure that you *have* to use it, because you may want to change the protective mount or the camera at some point in the future.

You can go out and buy outdoor cameras if you really want to, but it's not a requirement. Indoor cameras will work just as well if you create a protective mount that will keep the weather off the camera and away from the wiring.

Moving On

Now you can watch your birds from the comfort of your easy chair without having to even get up. Just flip the channel to have your own reality show. And you'll see far more than you've ever seen in the past, because you can watch the birds from a closer distance without disturbing them.

Project 16

No Keys Necessary: Installing a Keyless Entry

Equipment Needed

- AK Remote Control Door Handle (by Morning Industry)
- Screwdriver
- Drill with 1 inch bit

When keyless entry systems for cars first came out, they were all the rage. After all, what's better than being able to point a remote at your car and have it lock, even if you're half-way across the parking lot? Today, most vehicles have remote entry capabilities as a standard feature.

If only your front door had the same feature. You wouldn't have to struggle with an arm full of groceries or a struggling child as you try to unlock the door. And you definitely wouldn't have to worry about finding the keyhole after dark. (Let's not discuss why you haven't automated those front porch lights yet.)

Keyless entry for the home is quickly catching on as a great way to add a little security and convenience to your home. And you don't have to have a professional installer to have a keyless entry system installed. In fact, you can probably pick up a keyless entry lock and handle set at most home improvement stores.

For this project, we're going to use the AK Remote Control Door Handle by Morning Industries. You should be able to install and program it in under an hour.

Installing the Lock and Knob

Note

Installing a new door handle and lock set is a fairly simple task under most conditions. However, for this project, we're going to assume that you're replacing an existing lock and knob. If you're installing a brand new one, you may need to cut a hole for the device, in which case you should refer to the manufacturer's instructions for the proper way to cut the hole.

1. The first thing you need to do to install the keyless entry set is to remove the old door knob. Open the door slightly and then remove the screws holding the handle together; there are probably two – one on each side. Then, grasp both door knobs and pull them apart. This may take a little muscle, especially if the door handle has been painted over a time or two.
2. Next insert the latch into the hole in the door from the side edge, as Figure 16-1 illustrates.
3. After you've installed the latch, then insert the exterior handle through the hole, lining it up properly with the latch.
4. Then slide the interior mounting plate into place and fasten to the door with the screws provided.

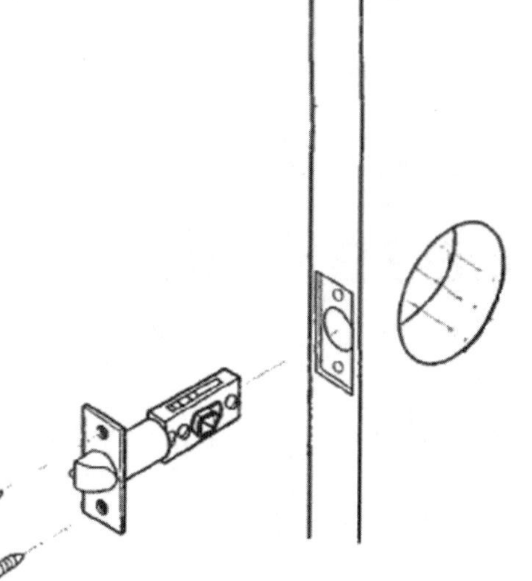

Figure 16-1 *Install the latch in the handle hole first.*

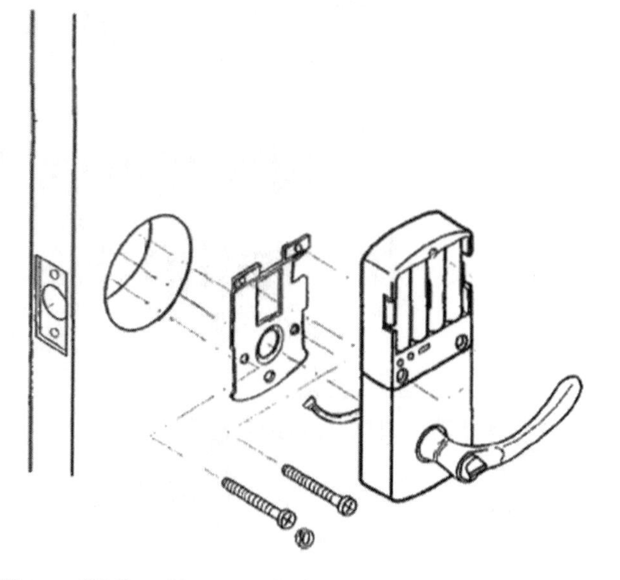

Figure 16-2 *First attach the interior mounting plate and then mount the interior knob.*

5. As Figure 16-2 shows, the interior handle should be mounted over the interior mounting plate.

When you have the handle and lock set put in place, turn the handle to make sure it is working properly.

Installing the Strike Plate

Now, if luck is with you, the existing strike plate will match up well with the new door handle. In some cases, however, you just won't get this lucky. If that's the case (and you can tell when you close the door if it's going to match up right or not), then you'll need to install a new strike plate.

1. To measure for the new strike plate, close the door until the latch touches the door jamb.
2. At the center of the latch, make a mark on the jamb, then open the door and measure toward the door stop, half the thickness of the door.
3. Then vertically mark the center point.
4. On that center point, drill a 1 inch hole about 1/2 inch deep.
5. Next, place the strike plate against the door and line it up with the center line you drew before drilling the hole.
6. Holding on to the strike plate, close the door to ensure a proper fit.
7. If the fit is good, then screw the strike plate into place.

Once you get the strike plate in place, then the only thing you have left to do is program the lock and remotes to allow entry.

Programming the Lock and Remotes

One concern that many people have is that remote-controlled entry systems could be compromised by someone with a remote that matches the door. However, you can program the lock and remote to work together, which means that unless you program a remote for someone else, they can't control your lock. And in the event that you lose a remote, you can reprogram the lock (and a new remote) so that it won't work with the old one.

1. To program your remote, you must first remove the battery cover from the locking mechanism. The cover is held on by a single screw at the top, center of the device.
2. With the battery cover off you can access the programming controls. Press the S button on the lock. You should hear a single beep.

3. Then, hold the remote in front of the lock, pointed at it, and press the remote button. You should hear two beeps that confirm the lock and the control are now synchronized.
4. You can repeat the synchronization process with up to 30 remote controls.
5. If you want to cancel remotes that have been previously programmed, press the C button. You should hear a single beep.
6. Then, using a **NEW** remote control, hold it in front of the lock, pointed at it, and press the remote button. You should hear two beeps, which confirms that **all other remotes have been erased and only this remote is synchronized to unlock the door**. This means you'll need to reprogram additional remotes if you need more than one.

Once your remotes are programmed, your keyless entry system should be working properly. Be sure to test the lock with each remote that you programmed, and have the manual override keys available in the event that a remote fails.

There is one more function of this particular remote locking system that might interest you. It's the ability for the lock to automatically relock itself 10 seconds after the unlock function is performed. To turn this option on or off, slide the Auto Relock button to the left or right. Left (marked as 1) turns the function on and right (marked as 2) turns the function off.

Understanding Remote Controls

It seems that these days everything is remote controlled. There was a time when your television or VCR were about the only things that you could control with remotes, but these days you can control everything from your home entertainment equipment to your fans, doors, and even starting your car.

In fact, according to some estimates, there are no less than four devices controlled by remote in any given home these days. Without a doubt it won't be long until we're controlling our entire homes by remote.

The first remotes (albeit very crude remotes) were designed in 1893 (U.S. Patent 613809). They've evolved considerably since that time, but they're still evolving as technologies change and improve. And even now there's more than one type of remote.

Remote controls operate by infrared signals, radio waves, and even Bluetooth technology is becoming popular. Infrared remotes are by far the most popular type of remotes in use today, though you'll also find some remotes of the other types as well.

Infrared remotes operate by passing an infrared signal to and receiving infrared signals from whatever device you choose to use. Infrared literally means "below red." The signals that pass between the remote and the device are light signals that are below the red spectrum, which is why the eye of your remote often looks red in color. You can't see this light, but if carries commands in the form of light waves.

Radio controls on the other hand use radio waves to transmit commands and Bluetooth remotes use Bluetooth technology to transmit commands. Most – if not all – home automation devices use infrared remotes. Which means that the remote needs to be in the line of sight of the device being controlled.

For example, if you're implementing remote controlled locks, opening the locks with the remote will require that you have a direct line of sight with the remote. That's because light travels in a straight line and anything impeding that line will block the signal. You may have experienced this with your television or stereo remote. If someone is standing between you and the device you're trying to control, the signal won't make it through, and so nothing happens.

With home automation, what you need to keep in mind is that most remote-controlled home automation devices use infrared signals. And most devices come with their own remote, however, a universal home automation remote that is also infrared should work with them.

In the movies you've seen the smooth guy that has a remote control that works with the lights, the stereo, and the fireplace? It's probably an infrared remote, and a little bit of light helps him set the scene for romance. If he can do it, why can't you?

Moving On

The next time you come strolling up to the front door with your hands full, all you'll have to do to unlock the door is press a button. Of course, that won't free your hands up to turn the handle to open the door, but it will make getting in the door a little easier. And for extra safety, you can tell the lock to relock automatically, which means you won't go to bed and forget to lock the doors ever again.

Project 17

Romantic Settings: Creating Grouped Tasks to Control Your Environment

Equipment Needed

- HAL2000
- Lamp Modules

How impressed would your spouse or significant other be if you invited them to dinner and then when they arrived you said something corny like, "It's time for a little romance?"

I'll bet they'd laugh at you and just go right on about their business. But if you said your corny line and the lights dimmed and soft music began to play, and the gas fireplace began to burn, you'd probably earn a lot of respect, or at the very least a little awe, right? Be prepared to be the person of the hour, then, because these are all things you can do when you create a group of tasks and trigger it with a voice command.

Of course, you'll need to use HAL2000 and you should have the appropriate lamp modules, appliance modules, and audio connections to make it happen. But by now you should have all of those in place, so we won't go into the minutia of installing all of that.

Instead, let's talk for a minute about giving HAL a voice command.

Speech Recognition

The HAL program can respond to your words with specific actions for two reasons. The first is that you've set the actions up ahead of time and have HAL programmed to perform those actions. But more importantly, HAL can respond because he can recognize the words that you're speaking. HAL has a built in speech recognition engine that works pretty hard to understand what you're saying.

There will be times, however, when HAL doesn't recognize what you're saying. At times like that HAL will ask you to repeat yourself, or it will send an audible sound – like a beep – to tell you your words were not clear enough.

To help HAL recognize what you're saying, be sure to speak clearly and pronounce your words properly. These actions will go a long way toward making HAL understand you better and to ensure that the actions that you *want* to happen are what actually takes place.

Setting the Scene for Romance

Now that you understand how important it is to speak clearly to HAL, let's get on with the romancing, shall we? In this project, we're going to set up a group of tasks that all happen together when you say a specific phrase to HAL.

Remember that you always need to say an attention word to queue HAL that it should be listing to the next words that you speak.

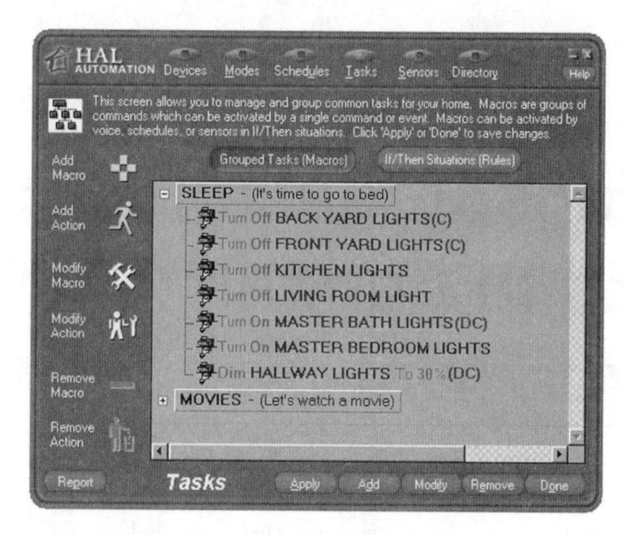

Figure 17-1 *The Tasks screen is where you set up grouped tasks and If/Then situations.*

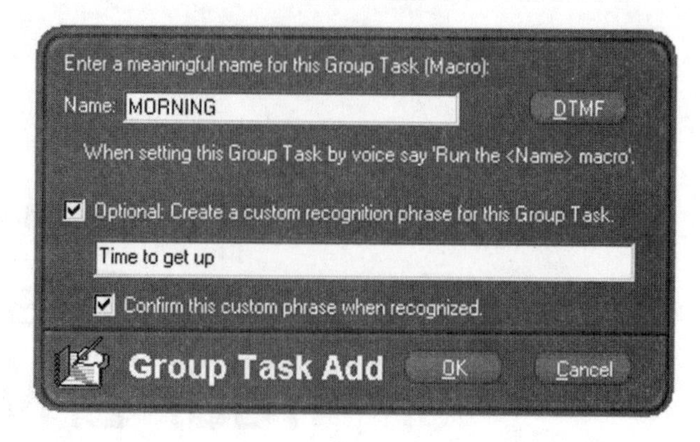

Figure 17-2 *The Automated Groups Tasks screen is where you begin creating your romantic setting.*

1. To start, right click the ear icon and then click on the Open Automation Setup Screen option. This will take you to the Automation Setup screen.
2. On that screen, click Tasks at the top of the screen, and you'll be taken to a screen like the one shown in Figure 17-1.
3. On the Tasks screen, click the Grouped Tasks (Macros) option and then click Add at the bottom of the screen.
4. You'll be taken to the Automated Grouped Tasks screen, shown in Figure 17-2.
5. In the Automated Grouped Tasks screen, enter a name for the group of tasks that you're creating.
6. Then click the box next to Optional: Create a custom recognition phrase for the grouped task, and enter the phrase you want to say to make this happen. In this case, we'll use the phrase, "It's time for a little romance."
7. You also need to decide if you want HAL to confirm the action when you say the phrase. If you select this option, when you say, "It's time for a little romance," HAL will respond to the phrase before taking the actions that you define.
8. When you're finished with your selections, click OK.

Adding Action to Your Group

Once you've clicked OK on the Automated Groups Tasks screen, you should be returned to the Tasks screen. From the Tasks screen, you can add the actions that you want to have take place when you say the recognition phrase.

1. From the Tasks screen, click the Add Actions option on the left side of the screen.
2. The Action Wizard will appear.
3. You've seen the Action Wizard many times by now. Use this Wizard to select the actions that you want to group together. In this case, let's select the Device option first and select the lighting that you want to be dimmed.
4. When you're finished with the settings for the lights, click OK and you'll be returned to the Tasks screen.
5. Click the Add Action option again, and this time, instead of adding a device, add music and select the desired playlist on the appropriate Action Wizard screen.
6. When you've made your selections, click OK.
7. Again you'll be returned to the Tasks screen.
8. Once again, click the Add Action option. This time select Send X-10 and in the screen that appears, set up a specific X-10 controller. Here we're going to make the assumption that you

have a gas fireplace and that you have the controls for that set up on an X-10 relay. If you don't, you can skip this step.

9. When you've finished making your selections, click OK to return to the Task screen.

Running the Action Group

You can continue to add actions to the group until every action that you have enabled and want to include is activated when you say your activation phrase. When you're done, however, you still have to run the Action Group that you've just created.

1. To run the group, click the Add Action option again. This time when the Action Wizard appears, select Run Grouped Task, as shown in Figure 17-3.
2. Select the grouped task that you want to run from the list that appears. Select the name of the task that you just created.
3. When you select the group, the actions that are included in the group will be displayed beneath it. Always check these actions to ensure they're ones you intend to happen.
4. Then select whether you want to delay the action or not and whether you want to confirm the action or not.
5. When you're finished, click OK.

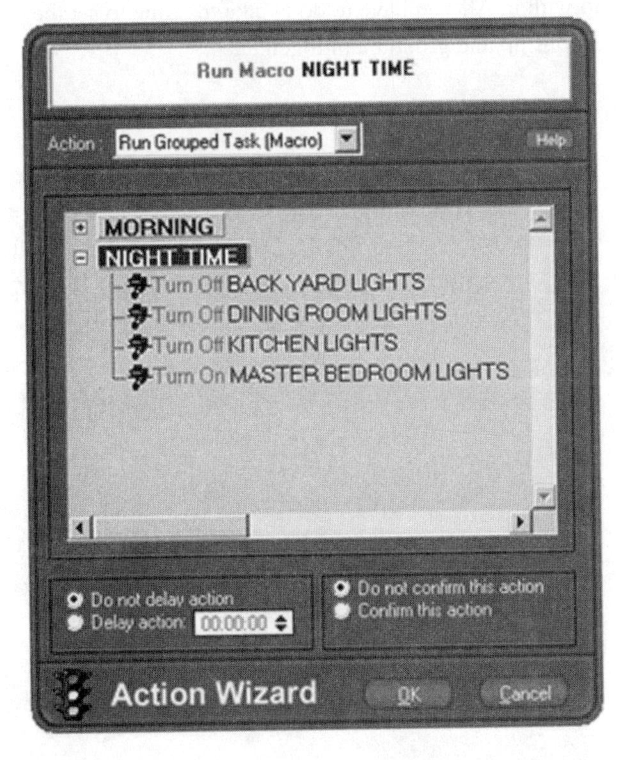

Figure 17-3 *To activate your grouped task, select Run Grouped Task from the Actions menu.*

You've just created and activated the group that's going to set the tone for your romantic evening.

Now when you want to run the group, all you have to do is get Hal's attention and then say, "It's time for a little romance," and all of the actions you just set will automatically take place.

Know Your X-10 Address

By now you've figured out that X-10 devices can work singly or they can work together as a group. What you might not have figured out is just exactly how they do that. How does a single device or a group of devices know when the signal traveling through the power lines in your home is meant for them?

The easy answer is the X-10 address. To identify devices, X-10 uses an addressing scheme that consists of letters and numbers. This scheme provides up to 256 unique addresses.

The first part of the X-10 address is the house code. This code is usually represented by any single letter from the range of A–P. Once you set a house code for your home automation system, all of the modules that you include in the system will have that house code.

Unit codes are the numerical portion of the X-10 address and are numbers in the range of 1 to 16. So, for example, a house code might be B12 or O1.

Each device, group of devices, or control module in your home should have its own device code, but the house code is shared. A single address for each group of devices allows you to control all of the devices with a single command. So, for example, if you set all of the lamps in a given room to a single X-10 address, then when you send a command for that address, all of the lamps in the room will respond in the same way.

Now, understand, most X-10 devices can only send or receive commands, not both. There are a few that can both send and receive commands. These devices are used in much the same manner as relay switches – commands that are received by a two way switch are then relayed to a one-way switch.

When you're considering home automation, take the time to think about *how* you want to control everything. Do you want to control all of the lights in one room, or only certain ones? Do you want to control lights *and* something else, or just the something else? How you want to control your devices should determine how you address them.

The tool with which you want to control the devices is also important. If you're controlling everything through your computer, then you have to worry less about the unit codes that you'll use than you might if you're using some remotes. That's because although you have 256 codes available to use, some controllers only allow a certain number of codes to be handled by that controller.

So, controlling your X-10 devices is a matter of codes, and controllers. It's still not a perfect system, but it has evolved considerably since X-10 was defined. Today, you can control far more than you've ever controlled in the past. And tomorrow you'll be able to control even more.

In the meantime, 256 devices is not very confining, especially when you consider that you can have multiple devices with the same X-10 code. And as technology changes and improves, more devices will be controllable in easier ways.

Moving On

You can set up grouped tasks for any number of different phrases or functions. For example, one phrase that you might like to program is "Let's watch a movie." Then you can set the phrase to turn on your television, dim your entertainment room lights, turn on the DVD player and adjust the volume of the speakers to the correct level. All you have to do is plug in the DVD and sit back and relax.

If you're really energetic, you can even program a situation for when the movie is over that returns all of the equipment that was set for the movie back to its pre-movie state simply by saying something like, "The movie is over now," or "Roll the credits."

You don't have to limit your action groupings to romance or dimming lights. There's much more you can do with it. All you have to do is take the time to set the actions up and group them together.

Project 18

Hello Dave: Creating Automation Personalization

Equipment Needed

- HAL2000

"Hello, Dave. I'm so happy you're home."

Believe it or not, that's how your house can greet you when you come home from work. Wouldn't it be nice to hear a friendly voice and the words "I'm so happy you're home," after a grueling day at work? It would certainly make your night go better, wouldn't it?

The HAL2000 program can be programmed to say a lot of different things, including greetings like this. You could even have HAL whisper "Goodnight," each night as you're going to bed if you like. It's all in what you tell HAL to do.

In this project, I'm going to show you how to program HAL to greet you when you enter the house at the end of a long day. You've already had a good bit of experience with HAL through this book, so we're not going into how to install or set up HAL. But you will see some information that you've seen in the past. This is just to be sure that we're all on the same page as we get this program set up. It's a multi-part project, and you don't want to miss anything or it won't work properly.

How It Works

First a little explanation. HAL has a capability called text-to-speech. This means that you can type a phrase that you want HAL to say, using a few parameters, and it will say it. There's a lot you can do with this ability, but before you can do anything you need to set a house mode with which to use the text-to-speech function.

A house mode is a group of actions that take place based on a given command. For example, you could tell HAL, "I'm going to bed now," and he would automatically turn off all the lights in the living room, kitchen, and den and turn on the light in the Master bathroom and the Master bedroom. He could simultaneously turn the heat down a little or the air conditioner up a little, and set the security alarm.

So, a house mode is a group of actions that take place in response to a specific command or phrase. Now, we're not going to do anything quite as elaborate as those examples used above, but before the chapter is out you'll learn how to create a house mode and have HAL speak to you when the house mode is triggered. It all begins with creating the house mode.

Creating a House Mode

Now that you understand a little more about how the house mode works, let's put together one that's not too difficult. Of course, you can always go back later and alter it, add to it, or remove functions if you don't want to have them included.

The first thing you need to do is open the Automation Setup screen, using the commands that you've used repeatedly in previous chapters.

1. When the program starts, click the mode button near the top and then select Whole House Mode, as shown in Figure 18-1.
2. At the bottom of the Whole House Mode screen, click the Add button. This will bring up the House Mode Add screen, shown in Figure 18-2.
3. First, assign a name to the mode, and then if you want to control this house mode by voice, enter a voice code for it. So, for example, in the house

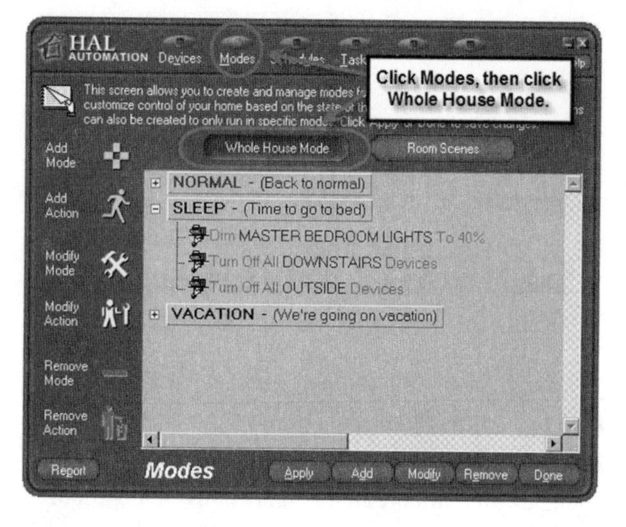

Figure 18-1 *The Whole House Mode is used to set up house mode actions.*

mode that we're creating, the phrase we might use would be, "Hi Hal. I'm home."

4. Finally, click if you want HAL to confirm what you've said. For this exercise, do not place a checkmark next to that option.
5. When you're finished, click OK.
6. Now, go back to the Modes screen, click the mode you just created, and then click Add Action to bring up the Action Wizard.
7. You've seen this action Wizard before, but we're going to go through it again.
8. When the Action Wizard screen, shown in Figure 18-3, appears, you'll need to select the various actions that you want HAL to perform. In this case, we're going to tell HAL to turn on all of the lights in the living room and Master bedroom at 60 percent.

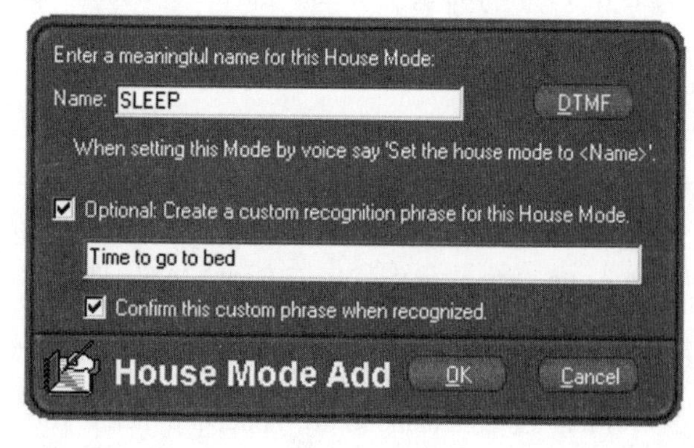

Figure 18-2 *Use the Add screen to begin a new house mode.*

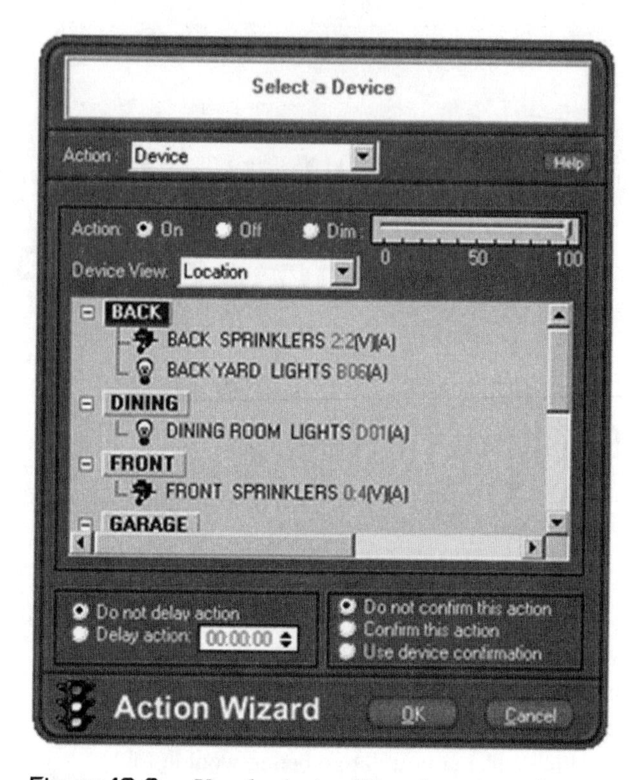

Figure 18-3 *Use the Action Wizard to set up the various actions that you want HAL to take when you announce, "Hi Hal, I'm home."*

9. To do that, select Device from the Action drop down menu. In most cases, Device is the default configuration, so you may not have to change anything at all.
10. Now, select Living Room to choose those settings and press OK. Then repeat the process with Master Bedroom.
11. When you click OK, you'll be taken back to the Modes screen. Click Add Action again, and the Action Wizard will reappear.
12. Now from the Action list select Digital Music Center. This will take you to the Digital Music Center Wizard shown in Figure 18-4.
13. On the Digital Music Center Wizard, select the music that you would like to have played, and be sure that the action is not set to delay and is not set for confirmation. When you're finished making your selections, click OK.
14. You're returned to the Modes screen. Click Add Action again and then click Device in the Actions Wizard a second time. This time, select the Master Bedroom from the device listing,

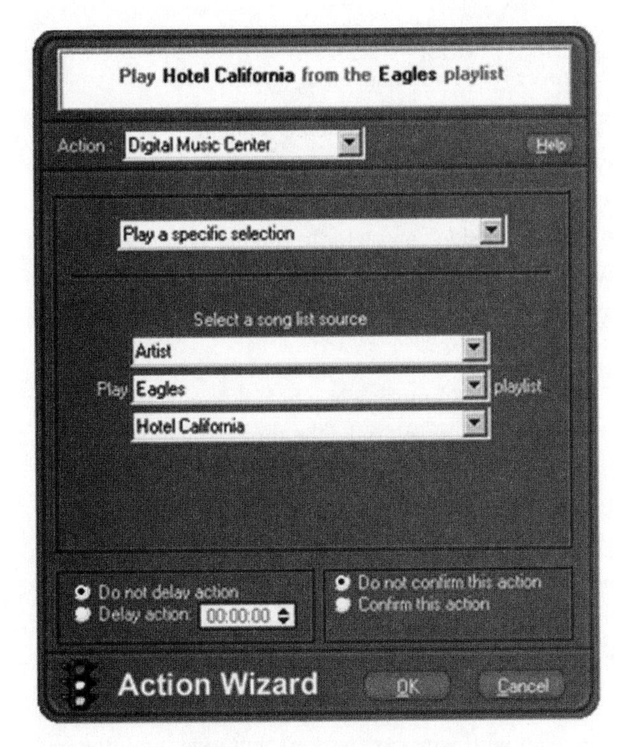

Figure 18-4 *The Digital Music Center Wizard helps you make music selections to include in your mode.*

adjust the settings to your liking and then click OK.

15. Finally, you're going to be returned to the Mode screen again. One more time, select Add Action, and then when the Action Wizard appears select Text To Speech. The Text-to-Speech Wizard screen appears.
16. On this screen, you'll find a text box where you can enter a phrase for HAL to say, as shown in Figure 18-5. Now, this is where it gets interesting.
17. HAL has a number of different voices and inflections that you can select from. Your user's manual lists all of them. For this project, we're going to use HAL's excited voice, which is designated by the character string <27>\Chr=excited\.
18. In the Text box, type: <27>\Chr=excited\Hi Dave. I'm so happy you're home.<27>\Chr=normal\.
19. The final character string at the end of the phrase is to tell HAL that after it says those two sentences, it should go back to using its normal voice.

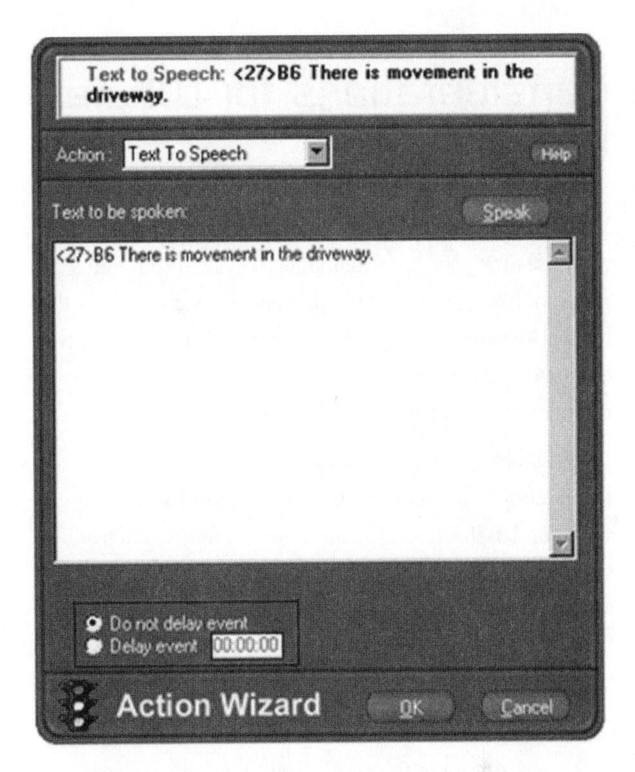

Figure 18-5 *Use this screen to enter the text you want HAL to say.*

20. Now, just to be sure that you've typed everything correctly, click the Speak button above the text box. HAL will say the words you've typed, in the tone you've indicated. If it doesn't, try re-typing the phrase.
21. One more choice you need to make on this screen is to choose not to delay the action. So, as soon as you finish speaking to HAL, it will respond to you.
22. When you've made the selection, click OK.
23. One final time you'll be returned to the Mode screen.
24. Now you've set up several actions to take place when you come in and announce, "Hi Hal. I'm Home."
25. Click Done on the Mode screen, and your mode will be saved and activated.

The next time you come in from work and use the announcement phrase, all of the lights selected and the music selected should come on. And HAL should say the phrase you trained him to say.

Considerations for Whole House Audio

When you're implementing voice controls or voice responses for your home automation system (through a program like HAL), you might want to be able to hear the responses in rooms of your home other than the room your computer is in. It's possible, but it does require that you run computer-connected whole house audio capabilities.

Whole house audio capabilities are not as difficult to implement today as they once were. You can still run wires under floors, in the attic, or through walls to connect audio throughout your whole house, but it's not always necessary. Wireless audio systems make piping audio into the bathroom easier than ever before.

There are a few things to be aware of when you're considering wireless audio systems. For example, most systems are not completely wireless. In most cases you'll find a receiver that still needs to be wired to your stereo or computer system. That receiver then connects wirelessly to the speakers that you want in your room.

If you do choose to go wireless, check the systems that you're considering carefully to be sure that you're getting exactly what you want.

Another consideration when implementing whole house audio, whether it's wired or wireless, is the requirements that will be placed on your computer (if you're connecting the system to the computer). Without the computer connection, you can still have whole house audio, but you will not have voice automation capabilities, such as hearing your voice automation program respond to the commands that you've spoken.

One more related issue with whole house audio and voice automation is the microphone that you're using to speak the commands to your computer. While having whole house audio is nice and will allow you to hear responses or sound effects from your voice automation software, if you don't have a whole house microphone system, then you won't be able to issue commands from whatever room you happen to be in.

Implementing housewide microphones can be an expensive and frustrating prospect. If this is the type of system that you'd like to have in your home, consider hiring an outside contractor to run the system for you. It won't be cheap, but in the long run, the quality of the system that you implement might be worth the extra expense.

If you plan to implement whole house audio or microphones yourself, take the time to plan it carefully before you begin drilling holes and running wires. Begin at the computer and audio connection and work your way out making sure that you know before you begin the placement of the speakers that you're running, the amount of cabling you'll need to accomplish the job, and any additional equipment that you'll need.

Whole house audio is a sweet project. It's not easy to run, even with wireless speakers, but give it a little more time and it will be as simple as connecting a wireless receiver and placing the speakers where ever you would like them to be. Until then, you can run a wired system or a hybrid system if it's important enough to you. Just be prepared for the expense and frustration of completing the system.

Moving On

So, now you know how to set Modes for HAL. They're a little complicated on the setup side, but the payoff is pretty big on the entertainment and usefulness side.

You can set up modes for any time of day that you wish – morning, bed time, when the kids get home from school. Or you can set them up for special instances, like coming home from work.

You're limited only by the modules and appliances you have automated and the creativity with which you program Modes.

Project 19

TV Anywhere: Wireless Television Throughout Your House

Equipment Needed

- Sony Location Free Base Station
- Sony Location Free Monitor
- Sony PSP
- Wireless Home Network

One of the greatest frustrations with television services today is the structure of their billing. For example, with some satellite television companies, you can have a special television receiver installed that will allow you to watch television on one to four separate televisions. But, that receiver is going to cost you extra money every month, and you have to pay a deposit on the receiver, and then you have to pay more for the service every month. And as if that weren't enough, you're going to have to sign a contract (anywhere from 12 to 18 months, depending on the company and the equipment).

Insulting.

In the ideal world, you would be able to pay for television one time and have it anywhere in your home, on any television. The problem is, we don't live in an ideal world. But we're getting close.

Wireless television transmitters, like the Sony Location Free system, are opening up options that promise to put satellite and cable television companies up against a wall soon.

I say soon because currently the technology is a little expensive. A starter kit that includes the base station and one monitor will set you back about $1200 at the time this book was written. But the price will come down in the future, and when it does, everyone will want one.

So we're going to get ahead of the game. This project is about installing and using the Sony Location Free system. Now, this system doesn't currently offer X-10 controls built in, so unless you connect it via an X-10 device module, you won't be able to access it with a program like HAL. And we're not going to add those kind of controls in this project. Instead, we're going to focus on installing the system.

Installing the Software

The Sony Location Free TV Pack (model number LF-PK1) comes with both a base station and a monitor, though you can purchase each piece separately if you prefer. When you install the system, you should start with the software, because the system requires a wireless network in your home to operate properly.

1. To install the software, insert the CD in your computer's CD drive.
2. The CD should auto-run and the installation screen will appear. Click Next.
3. The End User License Agreement will appear. Read it, and then if you agree with it, click select I accept the terms of this agreement and click Next.
4. On the next screen you'll be prompted to enter the software key from the CD case. Enter that number and then click Next.
5. Select the installation folder on the next screen and then click Next.
6. One more screen will appear, and when you click Install, the software should install.

7. When the installation is complete, you'll see a screen that says the installation was successful. Click Finish, and you're done.

Installing the software is just the first step in setting up the Location Free TV system. Once you've installed the software, you need to connect the base station to your television or other home entertainment equipment and to the Internet.

Connecting the Base Station

When you unpack the base station, it's going to be in two parts: the base and the unit. Put the base and the unit together, then you can begin connecting the appropriate cables between the base station and your television (or other equipment).

1. If you have an antenna cable, connect that to the back of the base station as shown in Figure 19-1. If you have cable instead of an antenna, then connect it instead.
2. Then connect the base station to the television or your other home theater equipment using S-video cables and the A/V jacks on the back of the base station.
3. Then connect the IRBlaster provided with the unit to the IR port on the back of the base station.

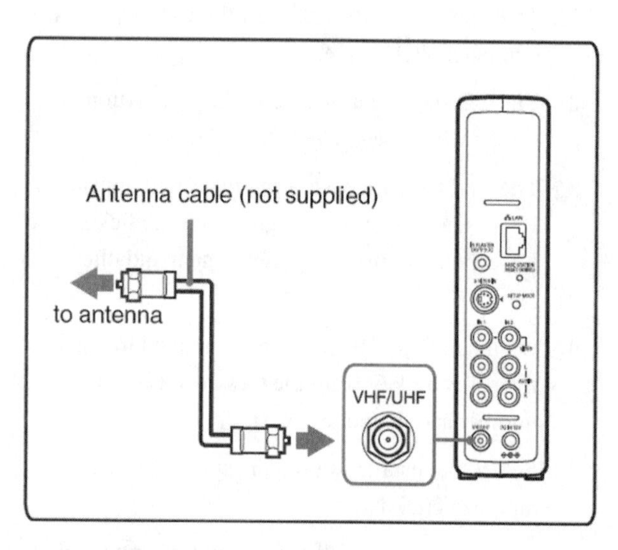

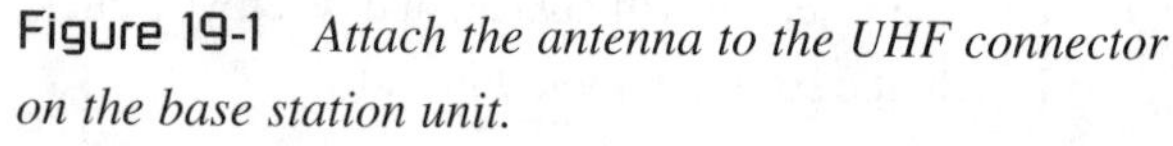

Figure 19-1 *Attach the antenna to the UHF connector on the base station unit.*

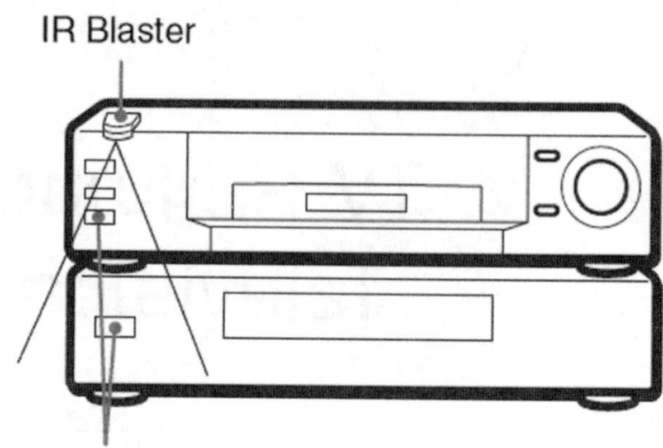

Figure 19-2 *Align the IR ports on the IRBlaster and your AV equipment to enable control by the viewing screen.*

4. The IRBlaster then needs to be placed within the line of sight of the IR controls on your television or home theater equipment, as shown in Figure 19-2.
5. Finally, connect the base station to your wireless DSL or cable router. You may need to give the base station security permissions for your network.

Once you have connected the base station to your audio visual equipment, you can sync it with the Location Free viewer.

Activating the Viewer

Now you can activate your Location Free viewer for the first time. Once you set the viewer up, you can carry it anywhere in your house so long as you have a signal from your wireless network.

1. Power on the Location Free viewer. Press and hold the Setup mode button on the back of the unit until the Set Up Mode LED light begins to blink.
2. Next, double click the Location Free Player icon on the desktop. The Location Free Player starts and the NetAV window appears. You don't need this screen for this setup, so close it and go back to the Base Station window.
3. In the Base Station Selection window, select the base station name that has the green icon next to it and then select Connect.
4. The Base Station will begin to register automatically.

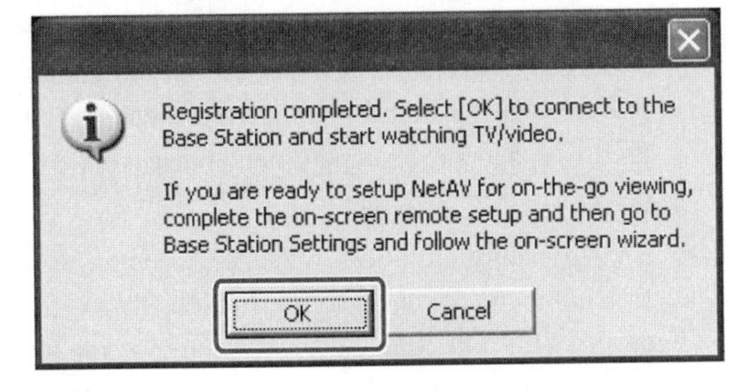

Figure 19-3 *When the registration process is complete, click OK.*

5. When the registration is complete, a message like the one shown in Figure 19-3 appears.
6. Click OK to restart the Base Station.

Setting Up Channels

Now all you need to do is set up the channels that the display unit will receive. Once you've set up channels, you'll have the same capabilities on the display unit that you have on your regular television.

1. To start the channel set up, click anywhere on the screen and the Index will appear.
2. Click the setting option on the bottom right corner of the screen.
3. The Settings Window will appear.
4. Click Channel Settings under TV/Video settings, and then select the channels that you watch from the list that appears.
5. When you're finished choosing your channels, click OK to enable the settings and close the window.

Now you're ready to watch TV on your viewing unit, and you should be able to turn it on and navigate through the channels without any difficulties at all.

Putting Location Free TV on Your PSP

One more option that might be of interest to you is the ability to use your PSP (Play Station Portable) to access Location Free TV, too.

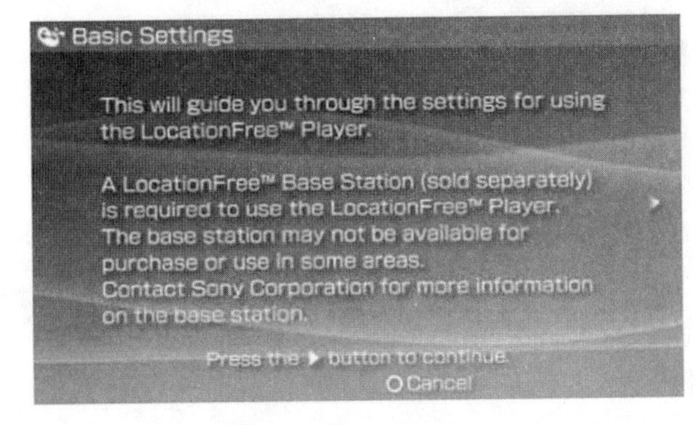

Figure 19-4 *The Location Free Player is already installed on your PSP, all you have to do is set it up.*

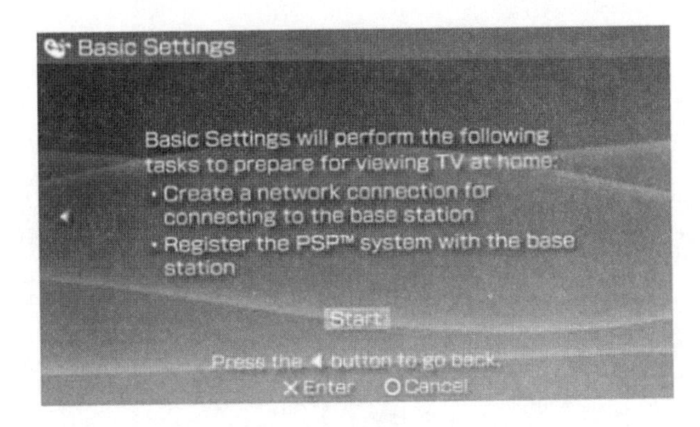

Figure 19-5 *The PSP will search for and register with the Base Station after you press the X button.*

1. To use the Location Free Player installed on the PSP, turn the system on and then turn on the Wireless LAN switch on the side of the PSP.
2. From the Home menu, select the Location Free icon and then the network icon.
3. The setting screen for the Location Free Player is displayed, as shown in Figure 19-4. Press the right arrow button to move to the next screen.
4. On the next screen, select Start and then press the X button. This starts the registration process with the Base Station, as shown in Figure 19-5.
5. Press and hold the Mode Setup button on the rear of the Base Station until the Setup LED light begins to blink.
6. Allow the Base Station to complete the setup process. This could take a few minutes, so be patient.
7. When the setup is complete, press the X button to finish defining the settings.

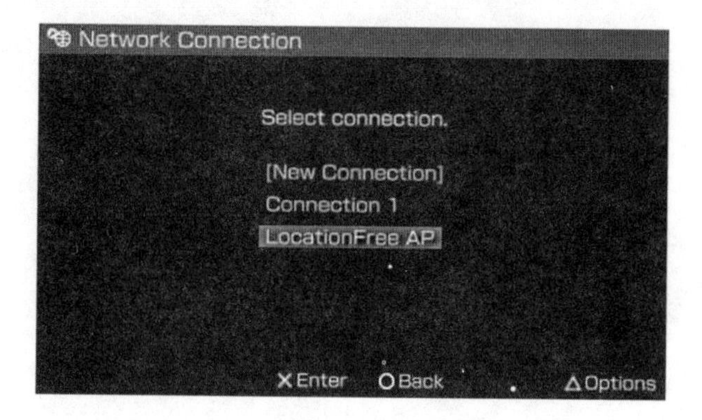

Figure 19-6 *Select the LocationFree button and then press A to connect to the LocationFree Base Station.*

8. Finally, connect to the Base Station by pressing the LocationFree AP button, as shown in Figure 19-6, and then pressing the A button. The TV screen will be displayed.

Now you not only have Location Free TV on the viewer that came with the Location Free package, but you also have it on your PSP. And if you have one of those, you know they have the most amazing display screen.

TV Anywhere. Literally

Have television all throughout your house without wires is a pretty neat trick. Especially when you consider the killing that cable companies are making charging extra money for each additional room that customers add to their cable service. The new technologies that allow you to have television anywhere in your home are also making other television capabilities possible.

During the later part of 2006, programs that began springing up make it possible to literally take your television programming with you. These programs allow you to download programming, down to specific shows that are stored on a special player which you can take anywhere you go. In other words, television fits into your lifestyle.

Here are some of the technologies that literally take your television anywhere:

- **Slingbox**: (http://www.slingmedia.com) This device actually lets you log into your home media connection to retrieve the television shows that you want to watch, no matter where you happen to be. Sling Media, the manufacturer of Slingbox, also developed a software program that can be used with handheld computers and other mobile devices to turn them into mobile televisions.
- **TV Anywhere**: (http://www.hauppage.com) The first in the mobile television revolution, TV anywhere is a PC USB2 device that allows you to connect to your computer-connected television using a high-speed Internet connection. The device even detects the resolution of the device that you're using to watch television remotely and compresses the video that's being transferred at an appropriate rate.
- **eMobile TV**: (http://www.pocketgear.com/software_detail.asp?id=22918) This is actually a software program that allows you to use your mobile device as a television anywhere you have a high-speed Internet connection. Unfortunately, this service is not the same as being able to connect to your subscription television service. Instead, this service offers a variety of television programming made available to eMobile subscribers.
- **Novac**: (http://www.novac.co.jp/products/hardware/vp-skype/nv-lf2000/index.html In Japanese). Novac is a Japanese company that offers an analog converter that, when used with the Skype telephone service, enables you to access and control your television from anywhere there is a high-speed Internet connection. Currently, this product is only available in Japan, but you can be sure that it will be available in other parts of the world in the near future if it's a successful product that sells well.

There are probably even more services such as these – Comcast offers one. And the limits of the service will be determined by the company that you choose to provide the device, software, or service.

So TV anywhere is more than just a possibility. It's a reality. And it truly will allow you to take your preferred television programming with you anywhere that you want to go. But it does require some additional investment in television equipment. It's not exactly home automation, but you have to admit, if television is your thing, then this is a very cool technology.

Moving On

See, there is a way to have television without having to pay outrageous provider fees to have it. Of course, right now the trade off is that you're paying Sony, instead, but eventually that will change, and Location Free TV or a technology like it will be much more mainstream and therefore more affordable.

Project 20

Adding Sound Effects to Your Home Automation

Equipment Needed

- HAL2000
- WAV Files (Your Choice)

Have you ever noticed that kids have sound effects for everything? If they do nothing more than turn around, they have a sound effect for it. And give a child a toy and they'll come up with sounds that you didn't know could come out of a child.

That's because a child's main job is to have fun. And sound effects are fun.

They would be fun to have in your automated house, too. For example, you could give your best friend a hard time when he comes over by having HAL say "Oh no, not you again!" when you say hello to him. Or, if your motion sensor that turns on a light in the hallway is tripped, you could have HAL play a snippet from your favorite song. You could even have the whole song played, if you like.

These are a few of the fun ways that you can use sound to add some fun and interest to your home. And that's what we're going to do in this project. We'll be using HAL and some WAV files to accomplish it.

What's a WAV File?

Throughout the book you've seen the term WAV file used a few times. What exactly is a WAV file? Well, obviously it's a sound file. But it's also a very specific type of sound file – a Windows and IBM waveform file. It is usually an uncompressed audio file. And because it's uncompressed, it is larger than an MP3 or some of the other types of audio file formats.

So, why is this important? Mostly because WAV is the only audio file format that HAL will play. Which means, as you're looking for audio files, you need to be sure the ones that you download are WAV files. The good news is that WAV files aren't difficult to find, and you can even get full songs from today's artists at places like MSN Music.

Creating Sound Effects

If you're really into the sound effects that are possible with HAL, you'll collect a variety of sounds over time. That's good. That means you have more to choose from when you're ready to create a sound effect.

Sound effects start with If/Then situations, because it takes a trigger for the sound to be played.

1. To set up an If/Then situation, right click the ear icon and select Open Automation Setup Screen.
2. Select the Tasks link at the top of the page, and then when the Tasks page opens, click the If/Then Situation (Rules) option.
3. At the bottom of the page, click Add to add a new rule.
4. The Rule Add dialog box appears. Type the name of the new rule into this box and then click OK.
5. The conditions screen, shown in Figure 20-1 will automatically appear.
6. Select an option from Trigger or Secondary. A Trigger is the primary cause of the rule to be activated while a Secondary is the secondary reason that a rule would be activated.
7. After you've made your selection, click Next.
8. Next, select the condition that you want to trigger the rule. You can use a device or even a date on the calendar.

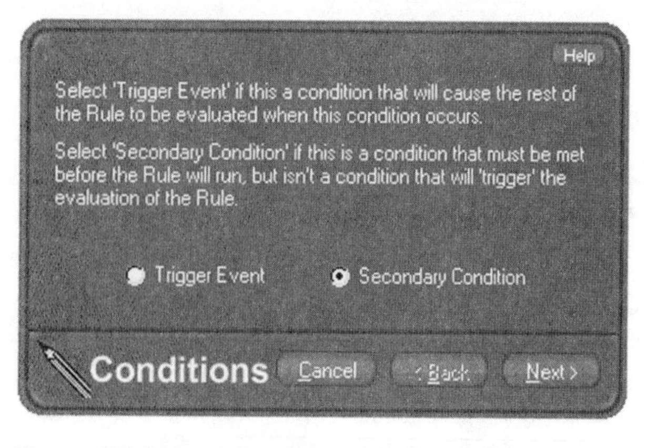

Figure 20-1 *The Conditions screen is the one you use to determine if the action is primary or secondary.*

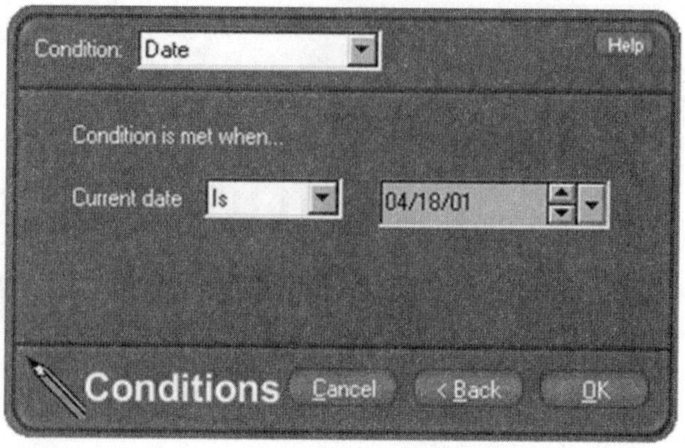

Figure 20-2 *The date condition lets you set an action based on a specific date.*

9. For example, if you use the date condition (shown in Figure 20-2), then you would tell HAL what date you want the sound effect to take place on. The same is true for days, or even specific times.
10. If you know that your spouse makes coffee at the exact same time every morning, you can set up a sound effect that is triggered by time that makes the sound, "Java, Java, JAVA," if you can find a WAV file that will do it. (And if not, you can always use Text-to-Speech to have HAL say it.)
11. So, select the condition that needs to be met for the event to take place. When you've finished your selection, click OK.
12. You'll be taken back to the Task screen. From there, click the Add Action button. The Action Wizard will appear.
13. In the Device drop down menu, select the Play WAV File option.
14. Then on the Action screen that appears (shown in Figure 20-3), use the drop down menu on the left to navigate to the location on your computer where the WAV file is located.
15. Select the file from the list on the right after you've found the location. If you need to be sure that you've selected the right file, you can click the Play button next to the Action field to preview the file.
16. After selecting the file, choose whether you want the action to be delayed or not, and then click OK.

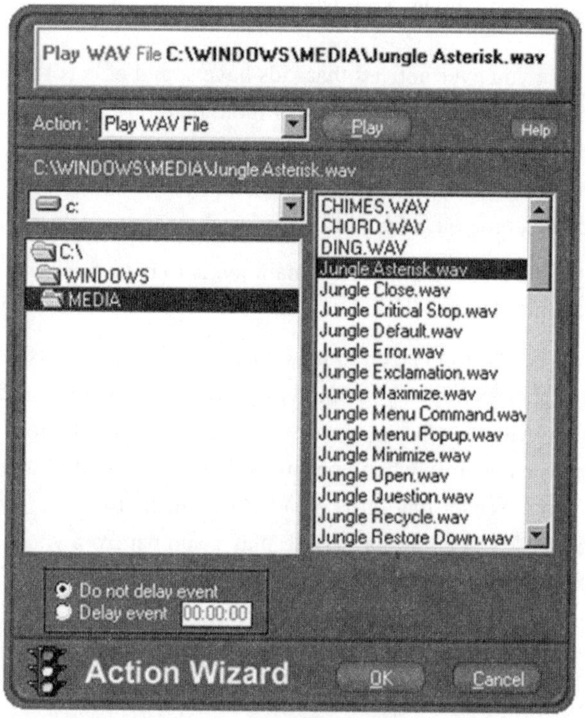

Figure 20-3 *Use the WAV File action screen to select the WAV file you want to play.*

Now you have an If/Then situation that specifies a sound effect and will be played when a certain action takes place. It's a neat way to add interest to your home automation system, and there's no reason you can't use it do a little showing off if you're having guests over.

You can always go back in and change or delete the scenario when it's no longer fun or useful, so have fun with it.

Putting Voice Activation to Work

Voice activation can be a fun addition to home automation projects. You can program bathroom scales and toilets seats or even your front door to be (comically) insulting. You can add holiday sound effects if that's what you enjoy doing. But voice activation isn't all fun.

Instead, the technology can be used to help people with disabilities, elderly adults, and anyone else who has trouble performing the daily activities that some people find so common place.

Take for example the movie, *The Bone Collector*, in which Denzel Washington played a quadriplegic detective. If you've seen the movie, then you know that Denzel (who plays Lincoln Rhyme) can control many devices around his home with voice commands.

These are not the kind of voice-activated abilities that are covered in other places in this book, but if you have someone in your home that could use a little extra help, they might be the kind of projects that you want to consider.

Here's an example. You can program your computer to pick up the phone, dial a specific number, and then broadcast the conversation over the computer speakers. This is useful for someone with limited mobility in their hands.

Or, you can program your home automation system to respond to your elderly father, who's living with you, when he tells the computer to turn on the lights or heat the oven. These abilities make life much easier for those with limited mobility or range of motion.

So, how do you go about doing this? For the most part, it requires setting up HAL or some program like it, to respond to voice commands. Then you need to install the appropriate X-10 modules, and before long, you'll have a wide range of voice control abilities in your home.

Even if you just want to control your home yourself, without any extraneous reason (such as limited range of motion) voice-activated commands aren't that difficult to perform. And with the right settings and abilities, you can even have the voice-activation software that you choose respond to you (or whoever happens to be speaking commands to the computer).

The one thing that you might want to do is begin building a sound library. You'll get tired of the handful of built in sounds that are included with the software program. Fortunately, programs like HAL offer you the ability to choose from sound effects that are on your computer's hard drive instead of just the sounds that are built into the program. For that reason, you might want to be on the lookout for sound effects that you like or that you think would be appropriate for your home automation system.

You can usually search the web for wav files that are compatible with most voice-activated home automation files. Just be careful that when you find a file that you'd like to download to your computer, the file isn't copyright protected. You don't want to get yourself in hot water unless you tell your voice activated system to heat the water for you.

Moving On

Home automation doesn't have to be all serious and technical. You can use your home automation gifts to add fun and interest to your life, and if you get a good laugh out of it in the process, so much the better. HAL is a good option for having fun with what you're doing, but there are plenty of other home automation management programs, too, and some of them may have equally as good or better planning and event creation tools. Use them. And have fun.

Project 21

Did I Close That Door? Installing an Automatic Garage Door Closer

Equipment Needed

- Autocloser by Xceltronixs
- Drill
- Screwdriver
- Ladder
- Pliers
- Wire Strippers
- Adjustable Wrench

It is so easy to get distracted as you're trying to get out the door in the morning. You're running late and your spouse wants a kiss, the kids want hugs, and the dog thinks he should go with you. By the time you make it to the car it's a wonder you can even think straight, much less remember to put the garage door down as you leave.

And unfortunately, If you leave the door up, you leave yourself exposed to anyone that wants to go in there poking around. If the door would just close automatically when you leave, then you wouldn't have anything to worry about, right?

Good news. It can. There's a gadget called the Autocloser by Xceltronixs. It works via IR sensors, and if the garage door sits open for a certain amount of time without the sensors being tripped, the Autocloser will automatically close your garage door for you. How's that for handy?

We'll focus on installing this handy little gadget in this chapter. Unfortunately, it's not X-10 capable, so the usefulness of the gadget ends at the closed garage door, but it's still nice to know that someone's got your back (door).

Installing the Autocloser

Before you do anything else, open the garage door. Then, as with any installation project that involves electricity, you should turn off the power to your garage door opener at the breaker box. Once you've turned the power off, use a circuit tester to ensure that you got the right breaker.

1. With the power off, the first thing you should do is bolt the reflector that was provided in the package to the J-brace on the garage door opener.
2. Next, apply the reflective tape included with the Autocloser to the reflector plate you just bolted into place.
3. Then connect the power to the Autocloser. The power outlet has a small tab on the top of it designed to help hold the power cord to the wall. To secure the tab, remove the screw from the center of the receptacle. Plug the adapter into the wall and thread the screw through the tab and back into the wall. Tighten the screw down, and now the adapter can't be accidentally pulled from the wall.
4. Now it's time to mount the Autocloser. It should be mounted with the infrared sensors aimed directly at the reflector plate and not more than 36 inches from the reflector plate when the garage door is completely open.

5. Now position the Autocloser device with the infrared sensor in line with the reflector plate. It will probably work best if the Autocloser is mounted on top of the garage door opener motor housing. The label on the Autocloser should be completely hidden when the Autocloser is mounted.
6. Locate the On/Off/Setup button on the Autocloser and switch it to the setup mode. If you have positioned the Autocloser correctly, it will emit a high-pitched beep to tell you that the infrared sensors are aligned and that the reflector plate is in the proper position.
7. Now you need to wire the Autocloser to the garage door opener wall button terminals. The terminals may be in different locations depending on the type of garage door opener you're using.
8. Next, wire the disable button to the garage door opener and then mount the disable, on a nearby wall.
9. Now you can test the Autocloser by switching the setup button to the on position and then opening, closing, and opening the garage door in rapid succession. Then, leave the garage door standing open.
10. The Autocloser is set to wait 5 minutes by default. After that 5 minutes it will close the garage door.

That's all there is to it. Now you don't have to spend half your drive to work worrying if you remembered to put the garage door down. The easy answer is it's down. Whether you put it down or the Autocloser caused it to close.

Common X-10 Problems

When you're setting up your home automation devices and applications, you might find that they don't always work the way you expect them to. That's when a good troubleshooting guide comes in handy.

This is a short guide of all of the most common problems that you might experience with X-10-enabled home automation projects. It's by no means comprehensive, but if you're in a bind, these are the most frequently experienced problems.

I've installed my X-10 devices according to the manufacturers directions, but they don't seem to be working.

There could be one of three problems with your X-10 devices or system. The first, of course, is that you either don't have the devices connected correctly, or you have them set to the wrong house and unit codes. These should always be the first things that you check.

Another problem could be "garbage" in your power lines. This is a problem that's especially troublesome in older homes where older wires have built up static and are old enough that they should be replaced anyway. If you're experiencing the garbage problem, then you can unplug all of the devices (except the X-10 device) in the room you're trying to automate.

Leave the power off for at least two minutes and then turn it back on. In most cases, this will give the garbage time to bleed out of your electrical wiring, and when you reconnect everything, you'll find that it's far more usable, and more successful.

The third problem that you might experience is just a difference in the band that the signals travel on. For example, X-10 devices use a couple of different bands (think of them as lanes on a super highway), and if your device is transmitting on one band, but your management program or receivers are on a different band, the signals will fly right past each other without ever meeting. To help solve this problem, buy a phase coupler and connect it to your computer. This should end the problem of everything transmitting on different bands.

I have installed X-10 modules and devices, but they aren't responding in the way that I would expect them to respond.

Occasionally when you do everything right, you'll still find that your X-10 modules and devices don't behave in the way you might expect them to. The devices being controlled by the X-10 might turn on or off unexpectedly. And it's all probably with good reason.

There are a couple of reasons that your X-10 modules and devices might not behave in the usual manner. The first is that there is some kind of interference in the line. The same electrical "garbage" that was mentioned above can affect the way that your X-10 modules and appliances behave. If you're having trouble with them, try unplugging them for a few minutes to allow the lines to clear.

If, when you plug your devices and modules back in, you're still experiencing problems, then it's possible that someone near you is also using X-10 home automation controls and you're both set on the same house frequency. Or, it's even possible that spikes in your electrical current are masquerading as X-10 commands. Try changing your house and unit codes. If that doesn't fix the problem, you may need to consider rewiring your home. Current spikes are a problem in older homes, which is why home automation is most recommended for newer homes.

I've connected all of my modules but they aren't working properly. What's the problem?

Modules that aren't responding are most frequently the result of mis-programmed house and unit codes. Try reprogramming the codes that allow the devices and modules to communicate and in most cases that will fix the problem. If it does not, refer to the questions above.

Moving On

Garage door openers and closers aren't the greatest home automation gadgets ever, but like keyless entry systems and some of the other projects that have been mentioned in this book, they can be handy little gadgets. And since it doesn't take long to install them, they're well worth the time and effort that go into them.

Project 22

Automating Your Hot Tub

Equipment Needed

- Leviton Receptacle Module
- Screwdriver
- Circuit Tester
- HAL2000

Several of my friends have hot tubs. And they love them, but there's one thing I've noticed about them that makes me a little nuts. When you decide you want a nice soak in the tub, you have to go out and turn it on and then wait 20 minutes while it's warming up. By the time the tub is ready, you no longer want to soak in it.

It would be so much better if the tub were completely ready to go when you were. Maybe your hot tub isn't set up that way, but you can fix that. All it takes is an X-10 compatible outlet. And you'll learn how to install that in this chapter.

> **Note**
>
> Some hot tubs are hard-wired into the house electrical wiring. Those require an X-10 compatible relay switch which installs simply. However, in this project, we're focusing on those tubs that have external (non hard-wired) power cords.

During this project, you won't actually be making any modifications to your hot tub at all. All you're going to do is remove the outlet where the hot tub is plugged into the power source and install one that is X-10 compatible. Of course, then you have to program HAL to turn the hot tub on at a specific time or when you say a certain command.

Figure 22-1 *The Leviton Wall Receptacle is X-10 compatible.*

The switch that we'll be using for this project is the Leviton Single Wall Receptacle Module, shown in Figure 22-1.

Installing the Switch

Before you can install the new wall outlet, you have to remove the old one.

Remember that you need to disconnect the power to the outlet at the breaker before you begin.

1. Once you've disconnected the power, unscrew the faceplate on the old receptacle and use the circuit tester to ensure there is no electrical current flowing through the wires.

Figure 22-2 *Rewire the new receptacle being sure to connect the wires properly.*

2. When you're sure the wires are not electrified, then you can disconnect the old fixture and dispose of it.
3. If necessary, you then need to strip the wiring to expose ¾ inch of bare copper at the end.
4. Then, connect the new receptacle to wires by twisting the wires together or using wire nuts to secure them.
5. The connections available on the receptacle are green, white, and black, as shown in Figure 22-2. Green is ground, white is neutral, and black is hot.
6. Once you've connected the wiring, seat the receptacle into the mounting box and secure it in place with the two screws that were supplied.
7. Before you place the faceplate on the outlet, use the small dials to set the house code. Then you can place the faceplate back over the receptacle and screw it into place.
8. Reconnect the power to the receptacle and then plug the hot tub into the outlet.

Automating the Tub

It's time to automate your hot tub, but you have a couple of choices on how to do it. If you use your hot tub at a specific time or on a specific schedule, then you can completely automate the hot tub by setting it on a schedule to automatically turn on.

Otherwise, you can set up a voice command that will trigger the automation. For this project, I'll walk you through setting up the automatic scheduling so that your hot tub turns on automatically on a set schedule. But before we get that far, you need to register the new X-10 device you just installed so that HAL can control it.

Registering the Device with HAL

1. To start the device registration process, right click on the ear icon and then select Open Automatic Setup screen. This should take you to the main Devices screen.
2. From the main Devices screen, click Add. You'll be taken to a screen where you are prompted to give HAL information about the device you're adding. Enter the requested information and then click Next.
3. On the next screen, you'll be asked to specify the address of the device you're adding. Do this by clicking the dials on the screen to set the house and unit codes. When the codes are set, click Next.
4. On the next screen, select the actions that you want taken when the device is active. For example, if you want the occupancy sensor to turn on specific lights when tripped, then select the option to turn on the lights.
5. On this screen, you'll also see an Options button. Click this button to define additional parameters – such as dimming options and two-way options. When you've finished with your selections, click OK and then click Next to move to the next screen in the setup sequence.
6. The next screen is where you set verbal confirmations from HAL. Remember that a

verbal confirmation from HAL can be something like, "I have turned the living room lights on," when you ask the program to turn the lights on, or HAL might confirm your request with "Would you like me to turn the living room lights on?" Set your preferences for confirmations and then click Next.

7. On the next screen you're prompted to set up groups for the device that you're adding. Unless you have a specific group that you would like to add the hot tub controls to, choose not to set up groups.
8. Once you've made your selections, click Finish, and the new device creation process will be complete.
9. Now you can program a regular schedule for your device by going to Action Wizard and then selecting the device and then the action that the device is to take.
10. You can then set up the schedule by going back to the Action Wizard and selection Calendar from the dropdown menu.
11. Make your calendar selections, and when you're finished, click OK. And you've waited on the hot tub for the last time. Your hot tub should now be set on your automation schedule, so it's ready when you are.

Controlling Appliances with X-10

Controlling appliances with X-10 is a bit of a tricky proposition. Some appliances can be controlled with external modules, others cannot. For example, most of the X-10 modules that you'll purchase in hardware stores and at home automation suppliers will only control devices that have on and off switches, not appliances like microwaves or ovens.

Don't be discouraged, however. If you'd like to use your home automation system to control your oven or microwave, you may be able to purchase the appliance with an X-10 module built into it.

Be prepared to spend a little extra money if you do decide to purchase these X-10-enabled appliances. The technology is more expensive to produce, and that expense is reflected in the price.

Another thing that you should know if you're planning to search out X-10-enabled devices is that some of them have their own software requirements, which means you may end up with two different management programs – one for the appliance, and one for the rest of the house. This is more of an annoyance than a problem, and you might be able to find a software that will combine everything, but at the very least you may be looking at changing your management application all the way around.

Appliances are a species all of their own when it comes to home automation. While it's very nice to be able to call your oven and preheat it while you're stuck in afternoon traffic, you may find that it's not worth the extra expense that you'll be paying to have that ability.

Moving On

One of the nicest things about home automation is that it works on your schedule. You can set up any action or group of actions and to take place when and how they suit you. So, when you're ready to get into your hot tub, it's ready for you. Or when you walk into the house, you're greeted with slowly increasing light and not pitch black darkness.

In the end, isn't that what home automation is all about? Having your house work on your schedule?

Project 23

Add a Little Humor to Your House

Equipment Needed

- HAL2000
- Humorous WAV Files
- Motion Detector

They say that laughter is the best medicine. The best medicine for what I'm not sure, but I have to agree that laughter will make a person feel much better. If you're having a bad day, if someone or something can make you laugh, the day just doesn't seem quite as bad, does it?

Unfortunately, there aren't always a lot of people or things around us that can usually make us laugh, so we have to take humor where we can get it. And fortunately, one of the places that you can have (or add) a little humor is in your home automation system. By adding some funny effects to your house, you can get a little giggle every single day if that's what you want.

In this project, we're going to set up a motion detector in the cabinet, so that every time someone opens the cabinet the HAL system says something like, "You're not eating again are you?" Or you might even prefer, "There's nothing different in here now that wasn't in here the last time you looked." Even a WAV file from the movie Little Shop of Horrors would be great: "Feed me, Seymour. Feed me."

Choosing a Motion Detector

For this project, you want to add a motion detector to your list of devices. I'd suggest a wireless motion detector, simply because you won't want to leave this little gag up for too long. Your spouse or significant other might have a real problem with hearing HAL every time he or she opens the cabinet door.

So, select a motion detector that you can move around. Then, when everyone gets bored with it you can change it up a bit. Maybe move it to your son's closet so that every time he opens the door he hears, "Help. Somebody please help me." He may not clean his closet out, but you can bet the first time he hears it, it will come as quite a surprise.

We've already covered how to install motion detectors and how to add devices to the HAL system, so I'm not going to bore you with the details of that portion of the project. If you need a refresher, you can look back to Chapters 3 and 22.

Somebody Let Me Out of Here

Finding the right sound to use with this little gag might take you a little while. And it's possible that you won't ever find exactly the one you're looking for. If you don't, then you could record your own. That will probably require some type of recording device. A digital voice recorder that will upload in WAV format would be an excellent choice.

If you don't want to go to the trouble of recording your own WAV files, then you could also use a Text-to-Speech script and have HAL say the phrase you want to use, rather than playing a WAV file.

If you opt to go that direction with your sound, then you'll need to change the action on the Action Wizard screen (in Step 13) from Play WAV File to Text-to-Speech.

Now, to put this little joke together, you're going to use many of the same steps that you've used in other voice-type projects:

1. Sound effects start with If/Then situations, because it takes a trigger for the sound to be played.

2. To set up an If/Then situation, right click the ear icon and select Open Automation Setup Screen.
3. Select the Tasks link at the top of the page, and then when the Tasks page opens, click the If/Then Situation (Rules) option.
4. At the bottom of the page, click Add to add a new rule.
5. The Rule Add dialog box appears. Type the name of the new rule into this box and then click OK.
6. The conditions screen will automatically appear.
7. Select an option from Trigger or Secondary. A Trigger is the primary cause of the rule to be activated while a Secondary is the secondary reason that a rule would be activated.
8. After you've made your selection, click Next.
9. Next, select the condition that you want to trigger the rule. You can use a device or even a date on the calendar.
10. In this case, the trigger will be the motion detector, so select the Sensor option and add the appropriate parameters.
11. When you've finished your selection, click OK.
12. You'll be taken back to the Task screen. From there, click the Add Action button. The Action Wizard will appear.
13. In the Device drop down menu, select the Play WAV File (or Text-To-Speech) option.
14. Then on the Action screen that appears, use the drop down menu on the left to navigate to the location on your computer where the WAV file is located.
15. Select the file from the list on the right after you've found the location. If you need to be sure that you've selected the right file, you can click the Play button next to the Action field to preview the file.
16. After selecting the file, choose whether you want the action to be delayed or not, and then click Ok.

Note

If you select the Text-To-Speech option, then you'll be taken to a slightly different screen where you can add the text, along with the voice effects, that you want HAL to say.

17. Now you have an If/Then situation that specifies a sound effect will be played when a certain action takes place.

This is an interesting way to spice life up a little, especially if you can set up the motion detector and the rules for actions when there's noone else around. It's much funnier to watch someone's reaction if they don't know what's coming.

Just have the video camera handy, because you might get lucky and record an America's Funniest Home Videos winner.

More About Text-to-Speech

Text-to-Speech, sometimes called Speech Synthesis, is an interesting program. It works by using software algorithms to convert printed materials to speech. The voice that's used is usually a computer generated voice, though in some of the more advanced program, you can synthesize a more human voice in either the male or female speech tones and patterns.

Text-to-Speech is also a lot of fun to use in home automation. With the right program, you can use text-to-speech to say funny lines or to tell you jokes when a certain event is triggered. Of course, if you don't change them frequently, the jokes will get old pretty quickly. The program doesn't have what it takes to make up (or think up) responses. Instead, it uses pre-determined responses, usually that you come up with.

The one real drawback to text-to-speech programs is that they're not all created equally. If you go to the Accessibility menu on your Windows-based computer, you'll find a link to various programs that are designed to help those with physical disabilities. One of those links is for the Narrator.

This is a text to speech program that's built into Windows. It's not a great program, but it will give you a general idea of how these programs work. If you find that the voice generated by this program is something that you don't want to hear regularly, then you should consider investing in a program (or an add-on module) that will convert text to a male or female voice.

Of course, in place of the text to speech option, you can use pre-recorded .wav files. The only problem there is that you lose the whole idea of converting text to speech. Instead, when you play the .wav files, you're getting a pre-recorded message or sound effect. It's not something that will read the text off a screen to you, like a text-to-speech program would.

Whichever option you decide to use – or a combination of options – there's more to your voice-controlled home automation system than just what's programmed into the computer. And that's one of the reasons that in previous chapters you were directed to create a sound library to draw from. This won't help with text-to-speech, but for voice response (to an action) you can find some files that are a lot of fun to use.

Moving On

Alan Alda (Hawkeye Pierce on M.A.S.H.) once said "When people are laughing, they're generally not killing each other." So, why not lighten up a little bit and give the folks around you something to laugh about for a few minutes. Using HAL and a few basic home automation tools, you can create some pretty funny pranks to share with your housemates, visitors, or even strangers walking down the street.

Just remember, it's fun, but don't take it too far. Even fun gets old after a while.

Project 24

Connecting to Your Favorite Entertainment

Equipment Needed

- HAL2000
- Internet Access

Home entertainment systems have been called the final frontier of home automation. That's because most often automated entertainment systems need to be installed by a professional. It's a difficult task to undertake for even the most experienced home automation guru.

Even so, that doesn't mean you can't use a home automation system like HAL to help a little with the whole entertainment conundrum. It's still a limited capability, but you can ask HAL for help when it comes to deciding what you're going to watch on TV on any given night.

In this project, we'll walk through setting HAL up to download television listings from the Internet. Then you can ask it, "What's on television at 8:00 tonight," and HAL can give you an intelligent answer (which is more than most television guides can give you at times).

Hooking HAL Up

For this project to work, HAL has to be Internet enabled. When you installed the software, you probably went through the Internet access set up. But just in case you didn't we'll give it quick coverage here.

1. If you didn't set up your Internet configuration when you installed HAL, you can do it now. Right click on the ear icon and then select Open System Settings.
2. In the Internet screen that appears (shown in Figure 24-1), first make sure that the Internet Enabled option is checked. This option must be enabled in order for HAL to access the television lists that you choose to download.

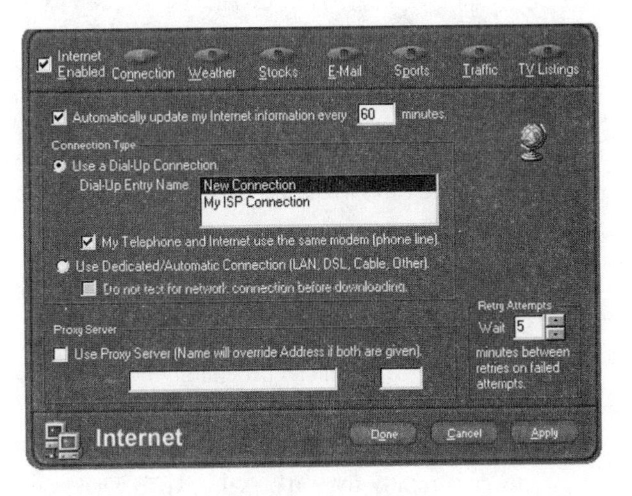

Figure 24-1 *Use the Internet screen to configure your Internet settings which will allow HAL to download certain types of programming.*

3. Next, configure your Internet connection using the options in the middle of the screen. If HAL is connected to the Internet already, these options may be filled out for you.
4. After you've made your selections, click Apply and your connection settings will be applied, but the Internet Window won't close.
5. Downloading Television Listings.
6. Currently, the Television Listings options in HAL are not as complete as you might like for them to be. HAL is an ever changing program, and as more Internet options for television become available, you'll have more choices. For now, however, you have to work with what's available.
7. To set up a new download schedule for TV listings, click the TV Listings button. You'll be taken to the screen shown in Figure 24-2.

Figure 24-2 *Use the TV Listings screen to select the channels for which you would like HAL to download programming.*

8. Scroll through the list of available channels and place a checkmark next to the ones that you want HAL to download programming information for. The maximum number of channels that you can download programming for is currently five, so be selective in what you choose.
9. You can also add channels that aren't including in the channel listing by selecting the Add button to the right of the channel listing.
10. When you click the Add button, the Channel Edit screen, shown in Figure 24-3 appears. Enter the channel code, number and name and then click OK. This won't guarantee that you can download the program listing for that channel. Many channels are still not available, but if it's available you'll be able to.
11. You'll be returned to the Internet screen. Select your time zone from the list below the TV listings, and then click Apply to stay in the Internet screen or Done.
12. Next, you'll want to go to the TV Listings screen to download the current programming. To do that, right click the ear icon and select View Internet Information.
13. The Internet Information screen will appear. Click TV listings to go to the TV Listings screen shown in Figure 24-4 and download and view current TV listings.

Figure 24-3 *Add channels to the possible listings available for download on the Channel Edit screen.*

Figure 24-4 *The TV listings screen shows current programming information.*

14. When you're finished with viewing the listing all you have to do is close out of it.

Now when you ask HAL what's on TV, it will be able to tell you. HAL will download a single day's worth of programming each time it connects to the Internet, so as long as it stays connected, or connects regularly, you'll have current program information available to you.

Windows, Mac, and Linux

When you're designing your home automation plan, one of the things that you should consider is the computer platform that you'll want to use with it. Obviously, if you're a Windows user you won't want to run out and buy a new Mac. It just doesn't make sense.

But that doesn't mean that all computer platforms are equal when you start talking home automation, either. For example, you'll find a lot of software for Windows, but the software that you'll find for Linux is far more customizable if you have the right knowledge.

The right knowledge is where most people find that they have problems. Writing programming script for a home automation system isn't the top of everyone's to-do list. What's more, it requires some specialized knowledge in either XML, C++, Perl, or some other programming language. If you don't already know these languages, it's probably not worth learning them, just to control your home automation system.

Still, Linux does seem to be the most usable of the operating systems when it comes to creating home automation that solves the issues that you're trying to solve. Mac is an excellent platform, and in most cases, you'll find that Mac-based home automation projects aren't too much different from Windows home automation projects. You'll find that Mac and Linux are somewhat easier to use if you know the language.

If you don't know all of the programming extras when you buy shoes, skip over the ones that require you to tweak the software to make it work. Instead, if Windows is available to you, use programs that are based on it. These programs are sufficient for many projects and they won't require that you learn anything new (assuming you're a Windows user).

Knowing which to use – Windows, Mac, or Linux – isn't easy; especially if you happen to be one of those people who happen to use dual booting machines or who have worked with all types of computers. You're a minority, however. By far, most people use Windows and they won't use any other protocol unless they're certain that they can get the support that they need. That support can be considerable if you don't know the programming languages that the programs are based upon.

Overall, Linux is the best, if you're a geek. If you're not, then Windows would be best. And no matter what platform you select, there will be some small issues that arise. It's all just part of using technology.

Moving On

HAL is still a ways away from being able to fully automate your home entertainment system. That's something that's best left up to professional home automation installers. But HAL can be useful as far as telling you what's on television at a given time. And more functionality is likely to come along soon, so for now, we just have to be satisfied with what we have available to us.

Creating an Illusion of Occupancy: Vacation Controls

Equipment Needed

- Lamp Modules
- Appliance Modules
- HAL2000

For many people, vacation is the only reason they keep going through the motions. Just to have that week or two away from work, from home, and from all of the usual hassle of daily life is enough to keep us going for months at a time. So, why would you want to ruin it worrying about your house and the security of it the whole time you're on vacation?

Most people wouldn't want to. The alternatives, though, aren't overly attractive. You can ask someone to house sit for you while you're gone. Or you can have friends and family drive by and check the house every so often. Or you could just automate some of the functions in the house, like the lights or television turning on, so that anyone who might be watching your house would believe that someone is there.

This project deals with setting up a vacation schedule for your home when you're not there. In the project, we're going to assume that you already have your lamps and appliances set up on X-10 capable controllers and that all you need to do is schedule the routine for while you're away.

If you don't already have devices and lamps connected to, and being managed by, HAL, then you need to set that up first. You can find more information about performing these installation tasks in many of the earlier chapters.

Setting Up a Schedule

To set up a schedule for automatic device control in HAL, you need to start in the Open Automation Setup

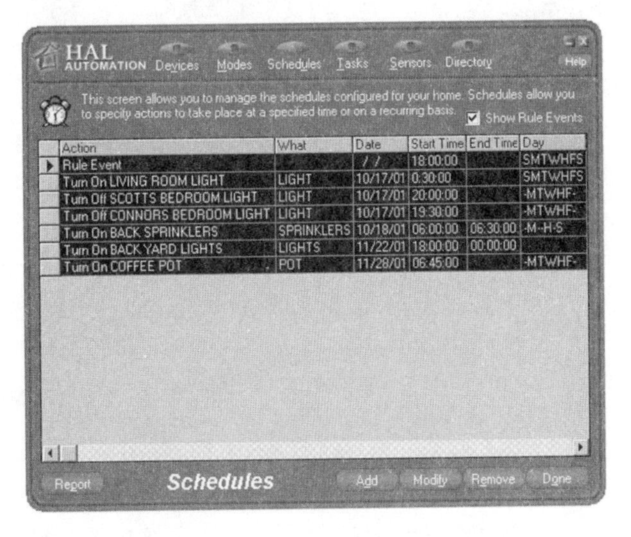

Figure 25-1 *The Schedule screen is where you set up schedules for events like vacations.*

Screen. Once you get there, click Schedules to go to the Schedules screen shown in Figure 25-1, and to set up the schedule that you want your house to keep when you're away.

1. To add events to the schedule, click Add at the bottom of the screen. This will bring up the Schedule Wizard, shown in Figure 25-2.
2. In the Schedule Wizard screen, select how often you want the event to occur.

Figure 25-2 *The Schedule Wizard walks you through creating a schedule.*

Figure 25-3 *Define the schedule on which you want events to happen.*

Figure 25-4 *Specify the time for each event.*

Figure 25-5 *Enter any recurrent times that you may want to assign to devices.*

3. On the Edit Action field, select the action that you would like to have occur. If you want to change an action, you can click Edit.
4. When you've made your Selections, click Next to advance to the screen shown in Figure 25-3. This screen is where you will specify days for the actions to take place. If you're going to be out of town, be sure to select dates that occur often enough to create an illusion of occupancy in your house.
5. It also helps if you vary the actions to some degree. Instead of just having the living room lights come on one night, you might have the living room lights and the front porch light come on, just so it will appear that there are people there or on their way back there.
6. When you've finished making your selection, click Next to continue on with the Wizard.
7. If you selected recurrent schedules, then your next screen will be to select when you want the occurrences to happen.
8. If you're not creating recurrent schedules, then you will go directly to the screen shown in Figure 25-4. This screen allows you to set specific times for each event.
9. Add the requested information and click Next.
10. If you are creating recurrent times, you'll be taken to the screen shown in Figure 25-5, where you're prompted to enter the recurrent time for each event.
11. Enter the recurring times that suit the schedule you've decided upon and then click Next.
12. If you've opted to set a schedule by Sunrise and Sunset, you'll be taken to the screen shown in Figure 25-6. On this screen, specify whether you want the event to happen before, after, or at sunrise or sunset. After you make your selections, click Next.
13. The next option you have is to select the house mode that you want to use, as shown in Figure 25-7. Since we're planning a vacations schedule, click the vacation mode, and then click Add to add it to the modes that will take place according to the schedule you have set.
14. When you've finished making your selections, click Finish and the event will be saved and the Wizard will close.

Now you have a vacation event that will make it appear as if your house is occupied, even if you stay in the Bahamas for the next three weeks.

Figure 25-6 *Set up schedules according to the sunrise and sunset times.*

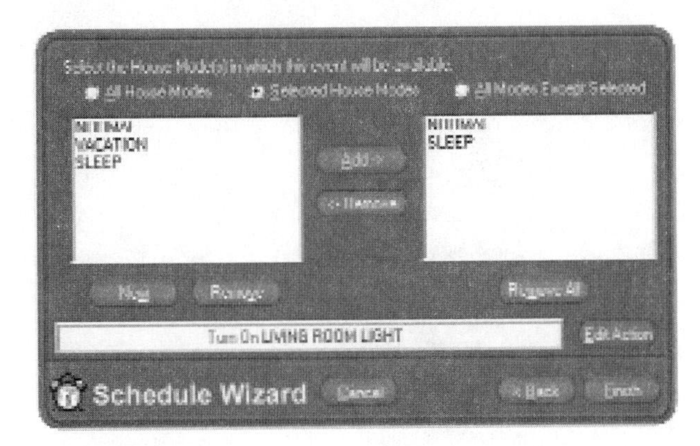

Figure 25-7 *Select the mode you want to take affect during the schedule you're creating.*

Other Safety Tips for When You're Traveling

Having a home that will function automatically when you're away is a great way to fool other people into thinking that you're still around. But a secured, automated home is no guarantee that nothing can go wrong.

So, in addition to the automated management applications that you need (which we discussed earlier in this chapter), you should consider adding some additional security processes. Yes, your house can manage itself, but what if it happens to experience problems?

Here's an example. Growing up, I knew a girl who came home from school one day and was doing her chores. After a few minutes, the girl started to notice a foul odor. It turned out that the house was on fire. No amount of home automation in the world can prevent those types of accidents.

For that reason, you should follow the suggestions below to ensure that your home is protected, no matter where you happen to be.

- Of course you'll want to lock up everything when you're not home. Even if you're using a video surveillance system, keep your doors locked. Thieves and criminals think to check out any entrance they can find.
- Make sure your home network is secure. Even if you have networked your home entertainment center into the network, you might want to double check that your network is secure. In today's broadband world, it takes only a few seconds for someone to break into your computer if it's not protected. Protecting it will slow the thieves and criminals down, because they're looking for any entrance into your system or network that they can find.
- Consider installing a monitored alarm system, or installing alarms on the system that you already have. Having your service monitored by someone else is just a safety precaution that ensures that someone else is also looking out for you.
- Never be too trusting. Installers, networking people, even people you've known for a long time can surprise you with the lies that they tell. Trust people, but not to the point that you're taken advantage of. Check references and business licenses, and make sure that there is nothing hidden in contracts that you can't take care of.

Security these days is a big deal. Everywhere you turn, someone is trying to sell you something that will make it that much more confusing when you're trying to stay protected. So, last words on how to find the security that you would like to have: ask those who already have security systems. Ask the people you're closest to for referrals, and always check references.

Moving On

Going on vacation doesn't mean you have to worry about your house sitting empty for the next week or two. It will be empty unless you have someone come by and check on it, but it won't appear that way to anyone who is watching your house.

You can set up a vacation schedule (or any other type of schedule, for that matter) that automatically turns lights on and off, plays music, and in general makes it appear that someone is there.

Anyone who might be interested in empty houses will walk right past yours, assuming that there's someone there.

Appendix A
Automation Checklist

Use this checklist to determine what elements of your home can be automated and what tools or equipment will be needed to complete that automation.

Automation

- ❑ Control Lights
- ❑ Control Switches
- ❑ Control Temperature/Environment
- ❑ Security Lights
- ❑ Security Cameras
- ❑ Voice Activation
- ❑ Control Appliances
- ❑ Monitor Children/Pets
- ❑ Monitor Wildlife
- ❑ Monitor Outdoor Activities
- ❑ Create Scenes
- ❑ Create Automated Actions

Tools and Equipment Needed

- ❑ X-10 Switches
- ❑ X-10 Modules
- ❑ HAL Software
- ❑ Management Software
- ❑ Drill
- ❑ Drill Bits
- ❑ Screwdrivers
- ❑ Wire Nuts
- ❑ Circuit Tester
- ❑ Computer
- ❑ Appropriate Networking Equipment
- ❑ Internet Access
- ❑ Telephone Lines/Service
- ❑ Masking/Electrical Tape

Note

You may need equipment or software that's not included on the list above. Use the blank space to the right of the list to add specific name brands or additional materials needed.

Circuit Box Labeling

Use this worksheet to correctly label your circuit box. Have a partner tell you what outlets/switches/fixtures are affected each time you turn one off, and then label that circuit next to the corresponding number below. Add additional lines if necessary. When complete, tape diagram to circuit box door for easy reference.

> **Note**
>
> Some circuits may require two lines because they use two circuit breakers that are connected.

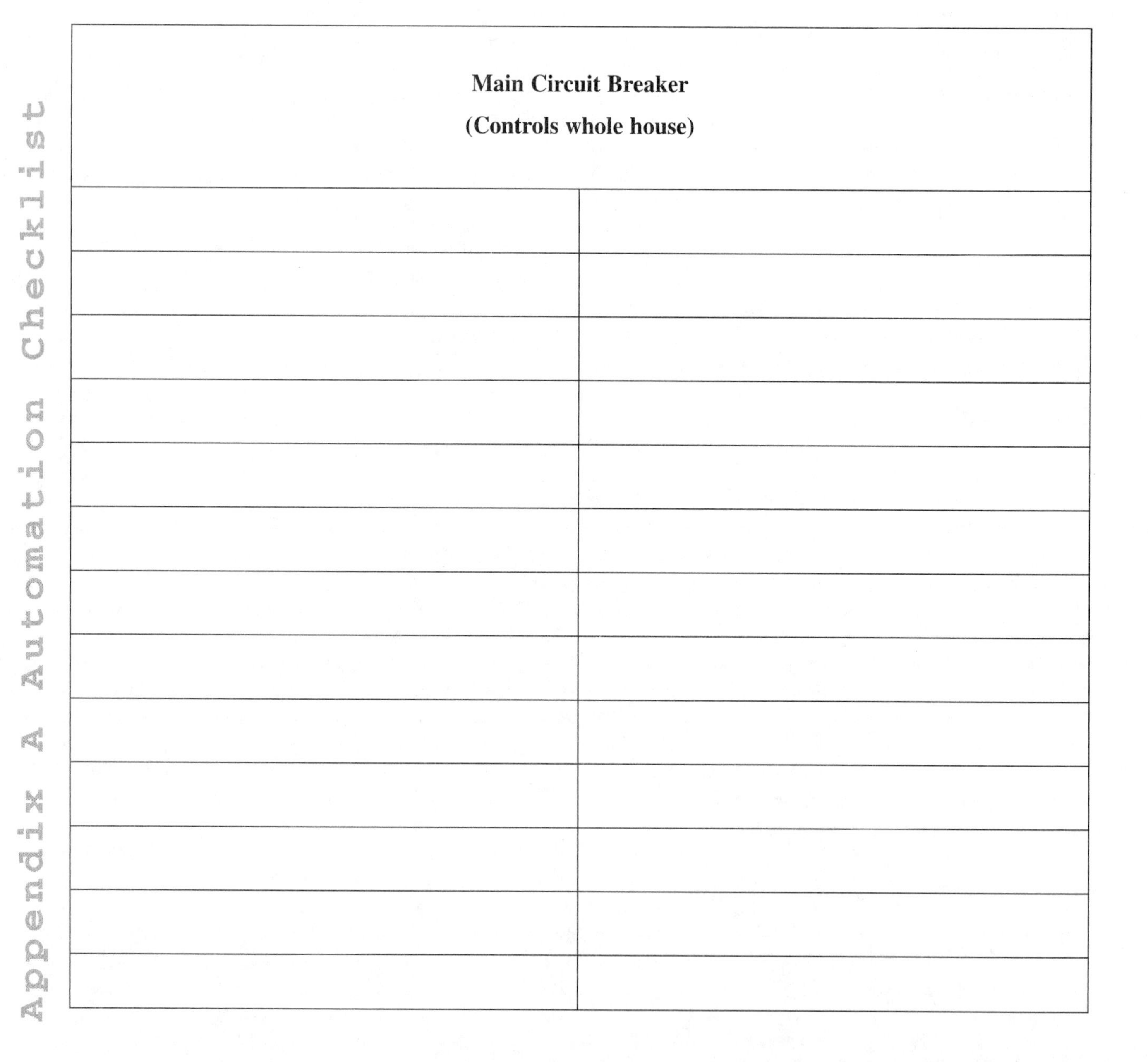

Main Circuit Breaker (Controls whole house)	

Outlet and Switch Diagram

One worksheet equals one room. Use this worksheet to diagram existing and new outlets and switches in each room. Mark existing switches and outlets in blue and new switches and outlets in red.

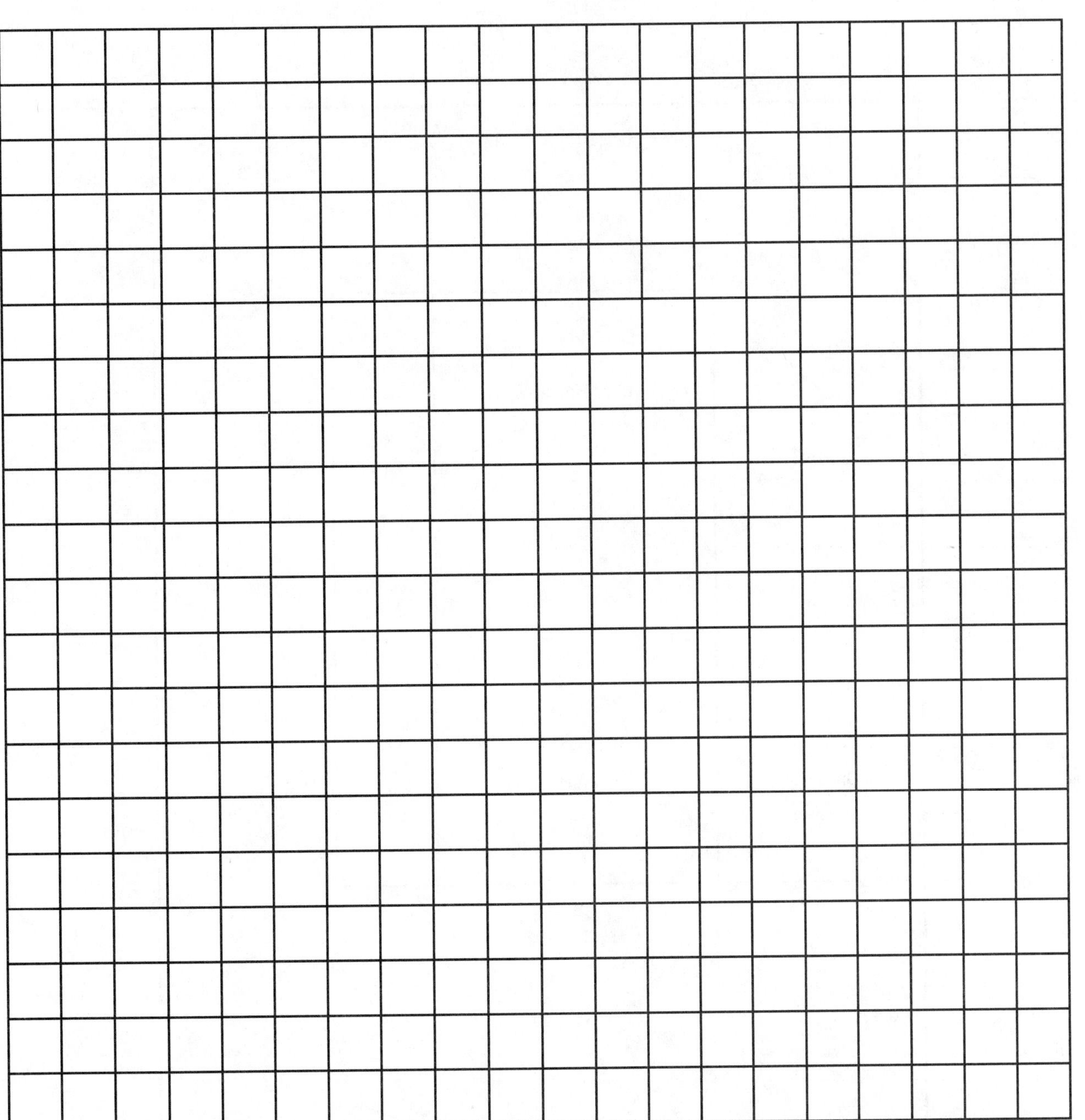

Outdoor Automation Diagram

Use this worksheet to diagram existing and new outdoor fixtures. Mark existing fixtures in blue, and new fixtures (cameras, motion sensor lights, etc.) in red. Be sure to make note of available wiring and power outlets if they exist.

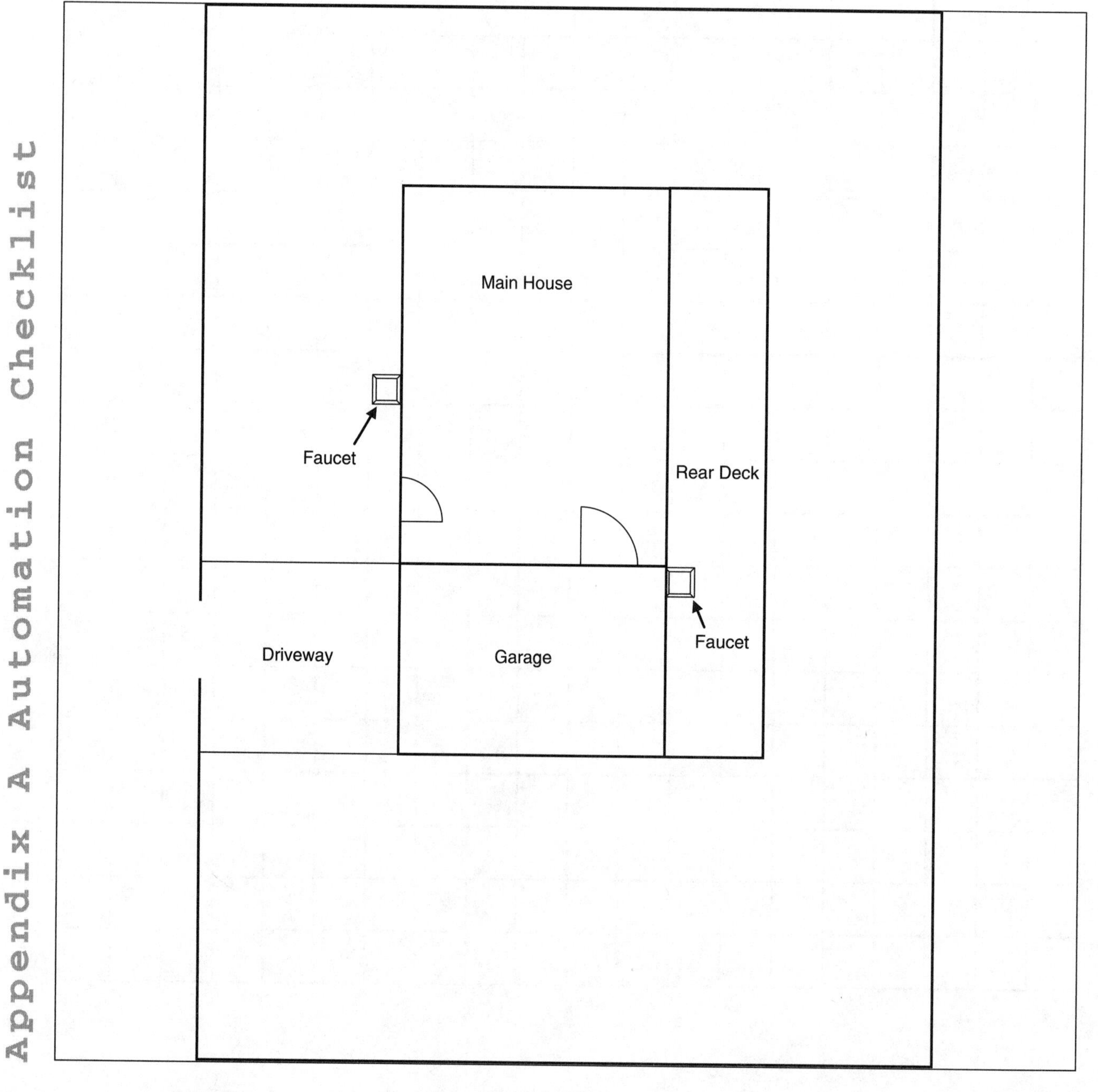

Appendix B Home Automation Resources

Use these online resources to learn more about home automation, wiring, specific home automation projects, and other aspects of home automation.

Information, Articles, and Tutorials

Beyond Logic
http://www.beyondlogic.org/

Bluetooth Homepage
http://www.bluetooth.com/bluetooth/

Circuit Cellar
http://www.circellar.com/

Convergent Living
http://www.coraccess.com/

Dan Lanciani's Home Automation Page
http://www.danlan.com/homeauto.html

DIY Home Automation (UK)
http://www.diyha.co.uk/

Do-It-Yourself Audio, Video and Infrared Remote Distribution Systems
http://sevinsky.20m.com/ha/
Doug Smith's Automation Page
http://www.smithsrus.com/HomeAuto/

Draconi Automation
http://www.draconidigital.com/draconiautomation/

EH Publishing
http://www.ehpub.com/

Electronic House Magazine
http://www.electronichouse.com/
Ennulat Home Automation System
http://www.homeautomationindex.com/he/hehasys.htm

FreeBSD
http://people.freebsd.org/~fsmp/HomeAuto/HomeAuto.html

FutureTech Home
http://www.futuretech-home.com/

HAI Automation Simplified
http://www.homeauto.com/main.asp

Hardware Book
http://www.blackdown.org/hwb/hwb.html

Home Automation Control Systems
http://www.hacs.com/

Home Automation FAQ
http://www.automationfaq.com/fom-serve/cache/1.html

Home Automation for the Mac
http://www.mousehouse.net/

HomeAutomation.Org
http://home-automation.org/

HomeAutomationSites.com
http://www.hasites.com/

Home Automation Wiring Guide
http://www.automatedhome.co.uk/contentid-2.html

Home Automation Wiring Guide
http://www.smarthome.com/howto13.html

HomeNet Help
http://www.homenethelp.com/

Home Networking and Home Automation Technologies
http://www.geocities.com/luisferm/

HomeTech Solutions' TechWiring System
http://www.hometech.com/techwire/index.html

HomeTech Solutions' Tutorials and Information
http://www.hometech.com/learn/index.html

How Bluetooth Works
http://www.howstuffworks.com/bluetooth.htm

HVAC/R Sites
http://www.elitesoft.com/sci.hvac/hvac.html

Intelectron Modification
http://www.homeautomationindex.com/intelectron.html

IntelliHome
http://www.intellahome.com/

IR Remote System
http://sevinsky.20m.com/ha/ir.html

Low Voltage Home Pre-Wire Guide
http://www.wildtracks.cihost.com/homewire/

Mister House
http://misterhouse.sourceforge.net/

Mobile Home Automation
http://www.lababidi.com/mobileAutomation/

Nuts and Volts Magazine
http://www.nutsvolts.com/

Project Guru
http://www.linuxguru.be/

Serial IR Controller
http://www.armory.com/~spcecdt/remote/

Structured Wiring How-To
http://www.swhowto.com/

Technick
http://www.technick.net/public/code/cp_dpage.php?aiocp_dp=_main

Tom Engdahl's Electronics Page
http://www.tkk.fi/Misc/Electronics/

What is X-10
http://www.smarthome.com/about_x10.html

Whole House Audio and Video
http://www.smarthome.com/wwaudvid.html

X-10 Home Automation Knowledge Base
http://www.geocities.com/ido_bartana/

X-Home.com
http://www.x-home.com/

X-10 Ideas
http://www.x10ideas.com/

Products and Services

Alarm System Store
http://www.alarmsystemstore.com/

Always Thinking Products
http://www.alwaysthinking.com/products/products.html

Amazing Gates
http://www.amazinggates.com/

Applied Digital, Inc.
http://www.appdig.com/irx.html

ASI Home
http://www.asihome.com/ASIshop/index.php

AudioAdvice
http://www.audioadvice.com/

Audioplex Technology
http://www.audioplex.com/

AudioSparx
http://www.audiosparx.com/sa/display/cat.cfm?soundamerica=true

Automated Environment Systems
http://www.automated1.com/

Automated Home Technologies
http://www.autohometech.com/

Automated Outlet
http://www.automatedoutlet.com/home.php

CentraLight Systems
http://www.centralite.com/

Channel Plus
http://www.channelplus.com/

Click Away Remotes
http://www.aclickawayremotes.com/

Concept Devices
http://www.concept-devices.com/

Control Plus
http://www.controlplus.co.nz/

Cortexa Technology
http://www.cortexatechnology.com/

CP290 Director
http://www.thewoodwards.us/sw/CP290Director/

CR Magnetics
http://www.crmagnetics.com/

Custom Automation Technologies, Inc.
http://www.customautomationtech.com/

DigitHouse (Australia)
http://www.digihouse.com.au/

Easy House Home Automation Project
http://www.jumbo.com/file/7273.htm

Elan Home Systems
http://www.elanhomesystems.com/

Electronic Solutions Company
http://www.electronicsolutionsco.com/home_
automation/

Electronics2You
http://electronics2you.com/

Evation.com
http://www.evation.com/irman/

GE Security
http://www.gesecurity.com/portal/site/GESecurity

Habitek (UK)
http://www.habitek.co.uk/

HeyU
http://heyu.tanj.com/heyu/index.html

HomeAutomationnet.com
http://www.homeautomationnet.com/
shopping/

Home Automated Living (HAL)
http://www.automatedliving.com/default.htm

Home Control Assistant
http://www.advancedquonsettech.com/

Home Electronic Ideas
http://store.homeelectronicideas.com/

HomeSeer
http://www.keware.com/

Home Team
http://www.hometeam.com/index.htm

Home Toys
http://www.hometoys.com/

HomeVision
http://www.csi3.com/homevis2.htm

HomeWatcher
http://www.homewatcher.com/

HouseBot
http://www.housebot.com/

iControl
http://www.icontrol.com/

iLight Intelligent Controls
http://www.ilight.co.uk/index.shtml

iLock Digital Lock System
http://www.cc-concepts.com/products/ilock/

Insteon
http://www.smarthome.com/prodindex.asp?catid=74

Intellicom Innovation
http://www.intellicom.se/

Jackson Systems
http://www.jacksonsystems.com/

JDS Technologies
http://www.jdstechnologies.com/

Jesco Electric Company
http://jescoelectric.com/

Landmark Sound Labs
http://www.landmarksoundlabs.com/

Leviton
http://www.leviton.com/

Lightolier Controls
http://www.lolcontrols.com/

LiteTouch Lighting Controls
http://www.litetouch.com/

Lutron
http://www.lutron.com/

Mac Home Automation
http://www.shed.com/

Magen Security & Home Automation
http://www.magen.ca/

MasterVoice
http://www.mastervoice.com/

Mile High Automation
http://www.milehighautomation.com/?gclid=CNCRprLHv4cCFSxMGgodJzhvHA

Multimedia Designs
http://www.multimediadesigns.com/

Niles Audio
http://www.nilesaudio.com/

On-Q/Legrand
http://www.onqlegrand.com/jahia/Jahia/pid/899

Orr Systems
http://www.orrsys.co.uk/electronics/index.html

PC Gadgets
http://www.pcgadgets.com/

Power Control Systems
http://www.pcslighting.com/

PowerDetector
http://home.nc.rr.com/tacairlift/PowerDetector/Powerdetector.htm

PowerHome
http://www.myx10.com/

PowerLine Communications
http://www.powerlinecommunications.net/

QSI Corporation
http://www.qsicorp.com/

Remote Central
http://www.remotecentral.com/

Residential Control Systems
http://www.resconsys.com/

Russound
http://www.russound.com/

SavoySoft
http://www.savoysoft.com/

Simply Automated
http://www.simply-automated.com/

SmartHome
http://www.smarthome.com/

SmartHomeUSA
http://www.smarthomeusa.com/

Space Home Concepts
http://www.spacehomeconcepts.com/

Sprinkler Controller
http://ourworld.compuserve.com/homepages/rciautomation/p6.htm

Stargate
http://www.imsamerica.com/stargate.html

SupervisionCam
http://www.supervisioncam.com/

TVSpider
http://www.vidiociti.com/

Universal Remote Control
http://www.universal-remote.com/

Vantage, Inc.
http://www.vantageinc.com/

Vaughan Smart House Systems
http://www.vaughan-systems.com/

Visual Home Commander
http://www.vhcommander.com/index.htm

Wireless Thermostat
http://ourworld.compuserve.com/homepages/
rciautomation/p1.htm

X-10.com
http://www.x10.com/homepage.htm

X-10 Superstore
http://www.x10-store.com/

Xantech
http://www.xantech.com/

ZenSys
http://www.zen-sys.com/

Zeus Home Control
http://www.zeushome.com/

Glossary

Action

Any action that is programmed to perform as part of a mode, schedule, or task. Actions are programmed in the User Interface screen.

Attention Word

A word or phrase used to put your automation into active listening mode. The attention word is spoken into the computer's microphone or a microphone that's part of a network.

Condition

A condition is used by itself or with other conditions to trigger a Rule.

Device

Hardware that can be connected to a home automation network in some manner so that the management software can control it.

DTMF

The initials stand for Dual-Tone MultiFrequency. This refers to the tones that a telephone generates when its buttons are pressed.

Environment

Refers to the ambient noise of the location from which a user is interacting with the home automation project. Speech parameters can be adjusted for each interaction type whether it's a microphone, local handset, or remote phone.

Flag

Flags can be used to trigger rules and the status of a flag can be set as part of an action. Flags are generally used as program "markers" that require an outside source to set the mark.

Macro (Grouped Tasks)

This is a series of actions that are carried out when your home automation program hears a single command or programmed phrase, or when specific numbers are pressed on a local or remote handset.

Microphone

This refers to any microphone that's plugged into the computer and that can be used as an input device for communicating with your home automation applications. The microphone could be a standard computer microphone that's connected directly to the computer or it could refer to a microphone that's part of a network.

Modem

The device a computer uses to transmit and receive data over the telephone line.

Output

In general output refers to anything that receives data of any type from home automation controllers such as an X-10 lamp module that receives commands to turn a light on or off. In most cases, the term "output" is used to refer to a specific piece of hardware that is connected to the home automation network.

Recognition Phrase

This is a phrase that can be used to launch a grouped task, house mode, or room scene. The recognition phrase is programmed when the task, house mode, or room scene is created or modified.

Relay

A relay is a type of output. Relays are contact closures – when a relay receives a command to power a device, the relay's contacts are connected so that an electrical current flows through it.

Rule

A rule refers to an If/Then statement that can be programmed to evaluate one or more conditions and to carry out a set of actions if all of the conditions of that rule are met.

Schedule

This term may be used differently in different home automation products, but in general it refers to the ability to have an action run automatically.

Sensor

Hardware that connects in some manner to home automation software and that conveys status changes as they occur.

Speech Recognition

This is a type of speech recognition where the program recognizes words that are spoken in a deliberate fashion.

Text-to-Speech

This term refers to a program's ability to talk by translating written words into sound. The words that are translated are part of the programs.

Text-to-Speech Script

A text-to-speech script is an action that can be run from a mode, schedule, or task.

Timers

Timers are countdown clocks that are created in the home automation software and that can be used in a rule.

Trigger

Any event, action, or change of status that starts a mode, schedule, or task. A motion sensor detecting movement could trigger a rule, for example, or saying a macro's recognition phrase could trigger that macro. Schedules are triggered when the computer's clock reaches a certain time.

Voice Recognition

This refers to the process of how some programs understand and respond to verbal commands.

WAV File

This is a type of audio format.

X-10 Technology

Technology that allows compatible controllers to send and receive signals over the standard power lines.

Index

A

B

C

D

E

F

G

H

I

J

K

L

M

T

U

V

W

X

Z